TURNBULL

The Plumbing Apprentice Handbook

The Plumbing Apprentice Handbook

R. Dodge Woodson

McGraw-Hill, Inc.

New York San Francisco Washington, D.C. Auckland Bogotá
Caracas Lisbon London Madrid Mexico City Milan
Montreal New Delhi San Juan Singapore
Sydney Tokyo Toronto

Library of Congress Cataloging-in-Publication Data

Woodson, R. Dodge (Roger Dodge), date.
The plumbing apprentice handbook / by R. Dodge Woodson.
p. cm.
Includes index.
ISBN 0-07-071772-9
1. Plumbing. 2. Plumbing—Vocational guidance. I. Title.
TH6123.W66 1994
696'.1—dc20 93-15862
CIP

1 2 3 4 5 6 7 8 9 0 DOC/DOC 9 9 8 7 6 5 4 3

ISBN 0-07-071772-9

The sponsoring editor for this book was Larry Hager, the editing supervisor was Nancy Young, and the production supervisor was Donald F. Schmidt. This book was set in Century Schoolbook. It was composed by McGraw-Hill's Professional Book Group composition unit.

Printed and bound by R. R. Donnelley & Sons Company.

This book is dedicated to the two most important people in my life, Afton and Kimberley.

ABOUT THE AUTHOR

R. Dodge Woodson is a licensed master plumber, plumbing contractor, general contractor, and licensed real estate broker. He is the author of many magazine articles, as well as eight books, including *Home Plumbing Illustrated*, and *National Plumbing Codes Handbook*.

Other book titles by R. Dodge Woodson are:
Profitably Managing Your Rental Properties
Rehab Your Way to Riches
The Complete Guide to Buying Your First Home
The Complete Guide to Home Plumbing Repairs & Replacements
Get the Most for Your Remodeling Dollar.

Contents

Introduction

Welcome to *The Plumbing Apprentice Handbook*. Whether you are just entering the plumbing trade, thinking of becoming a plumber, are a plumber, or just want to know more about plumbing, this book is for you.

The author, R. Dodge Woodson, is a master plumber with nearly 20 years of experience. He has worked with residential and commercial plumbing. His work has included new construction, remodeling, and repairs. In addition to being a seasoned master plumber, Dodge has also been an instructor of various plumbing courses at the Central Maine Technical College. These courses have included apprenticeship plumbing and plumbing code classes.

During Dodge's long plumbing career, he has trained hundreds of plumbers. As an author, he has written numerous books, many on the subject of plumbing. If you want to learn about plumbing, you couldn't ask for a better teacher.

What will you learn from this book? You will learn the truth about the plumbing trade. Detailed information covers theory, techniques, and real-world experiences—the kind not often taught in a classroom.

The chapters are laid out in a logical order of progression. You will start with Chapter 1 and learn how the apprenticeship process works. When you move on to Chapter 2, you will learn the value of field experience. Chapter 3 tells you about the plumbing code: why it exists and how it affects you.

Safety is the topic for Chapter 4, and Chapter 5 delves into the basics of plumbing and related science issues. As you flip through the pages of Chapter 6, you will see, from dozens of illustrations, the types of tools plumbers depend on.

Chapter 7 outlines the fundamentals for reading and understanding blueprints; it includes illustrations of the types of symbols you will find on working drawings. Math is a sore subject for some people, but in Chapter 8 Dodge shows you how to simplify the mathematical needs required for plumbing.

Residential plumbing is highlighted in Chapter 9. You will crawl in the truck with Dodge and go to work in this chapter. The sometimes complex world of commercial plumbing is described in Chapter 10. When you get to Chapter 11, you will learn everything you need to know about well systems, and Chapter 12 does for septic systems what Chapter 11 does for well systems.

As you move along through the other 14 chapters, you will learn about such topics as water conditioning systems, water services, sewers, underground plumbing, drainage systems, vent systems, water distribution systems, gas piping, fixtures, institutional plumbing, special plumbing requirements, testing procedures, maintenance, drain cleaning, and plumbing repairs. When you reach the end of the book you will find a helpful glossary.

Hundreds of illustrations accompany the text and make Dodge's easy-to-read text even easier to understand. Unless you are a master plumber, and maybe even if you are, you will find this book a wealth of helpful information.

There are several active plumbing codes in use across the United States. Many towns and cities adapt and amend a code to suit their particular needs. When you see examples of code requirements in this book, please understand they are being used only as illustrations. They are true code considerations, but they may not be accurate in your location. Always check with your local plumbing code before installing any plumbing.

The author has made every attempt to provide you with accurate, current information. However, mistakes are sometimes made and good books have long shelf lives. Sometimes material becomes outdated as laws and regulations change. Always consult your local code enforcement office prior to starting any work.

Now, if you are ready to embark on an enlightening journey into the world of the working plumber, please read on.

Acknowledgments

I would like to thank my parents, Maralou and Woody, for all the help they have given me over the years. They have been special parents and good friends.

Andrew Wallace and Fran Pagurko are thanked for their cooperation in modeling for various photos throughout this book.

The following companies are thanked for their contributions of artwork:

CR/PL, Inc. (Crane Plumbing Fixtures)

General Wire Spring Co.

Goulds Pumps, Inc.

Hellenbrand Water Conditioners

Moen, Inc.

Ridge Tool Co.

Thank you all for helping in the creation of this book.

Chapter

1

The Apprenticeship Process

The apprenticeship process for the plumbing trade is a means of training people to become efficient and proficient plumbers. This process normally takes up to 4 years and involves both classroom and on-the-job training. The apprenticeship process is required by many jurisdictions before a professional license is issued to individuals for the purpose of performing work as a licensed plumber in the trade.

It is not uncommon for local authorities to require any individual assuming any plumbing duties to register as an apprentice or licensed plumber. In the old days it was not uncommon for people to enter the plumbing trade as helpers. The helpers dug ditches, drilled holes, and gradually learned the plumbing trade. When the helpers were prepared, they could sit for an exam in their attempt to become licensed plumbers. This is how I became a plumber, but times have changed.

Today's code requirements generally demand that anyone doing any type of plumbing work be licensed. Helpers can still dig ditches and drill holes without being licensed, but they cannot solder pipes, glue joints, or put hangers on pipe. When I entered the trade, helpers could do as much plumbing as they were capable of, so long as they were under the direct supervision of a licensed plumber; this is not the case today.

There is no need to get discouraged; you can still learn plumbing the old-fashioned way, but you will probably have to register as an apprentice. This registration doesn't mean you must attend classes; it merely means that you are licensed to work as a plumber's helper. The fees for these apprentice licenses are minimal.

Before we jump into the details of classes and field requirements, let's look at the structure of the plumbing trade and licensing requirements. The first license obtained is the apprentice license. Normally, this license can be obtained without any special training or experience. The license entitles the holder to work in the field of plumbing, but the licensee must work under the direct supervision of a licensed plumber. This means that an apprentice cannot be left on the job to work while the plumber goes to the supply house or convenience store. Even if the work being done is menial, like installing pipe hangers, an apprentice may not do it if there is not a licensed plumber on the job.

The next step in the plumbing ladder is becoming licensed as a journeyman plumber. This goal can be achieved in various ways. Some apprenticeship programs will allow students who have successfully completed 4 years of classroom training to test for their journeyman licenses. Under these conditions it is possible for students to become licensed plumbers with minimal field experience.

There are apprentice programs that combine classroom courses and on-the-job training. This type of program might require 2 years of classroom instruction and 2 years of on-the-job experience before testing for a journeyman license is possible.

It is also possible in some cases to test for a journeyman license without classroom training. Typically, at least 2 years of field experience are required before this type of licensing is possible. Local jurisdictions make and interpret their own rules, regulations, and laws; therefore, it is necessary to check the requirements for licensing on your local level.

Once you have a journeyman license, you are a licensed plumber, but you are still restricted in what you can do and how you may do it. For example, journeyman plumbers are not normally allowed to apply for and receive plumbing permits. Within the trade, only a master plumber has the authority to be issued plumbing permits. Therefore, a journeyman plumber is very limited in the work that can be done independently.

Journeyman plumbers, for the most part, must work under the supervision of a master plumber. However, they are not required to work under the direct supervision of a master. This means that licensed journeymen can go out on jobs alone. They can install plumbing without direct supervision, and the master plumber can check their work periodically.

The highest rung on the plumbing ladder is that of the master plumber. If you want to have your own plumbing business, you will have to either be licensed as a master plumber or hire one to represent your company.

The amount of time necessary to make it to the level of master plumber will depend on the requirements of your local authorities. Typically, it will take at least 4 years to become eligible for taking the exam for a master plumber's license. Normally, a journeyman plumber must work at least 1 year as a journeyman before testing for the master's license. Depending upon local regulations and the apprenticeship programs available, it should take between 4 and 5 years to become eligible for taking the master's exam.

This may sound like a lot of effort, but I believe it is worthwhile. Would you like to make $38 per hour? You can make that much and more as a master plumber. Of course, to make that kind of money, you will be self-employed and working directly for consumers. As a plumber on the payroll of a company, you should make between $25,000 and $40,000 a year, depending on where you are working and what type of plumbing you are doing. The income potential for master plumbers is good.

Plumbers are vital in protecting the health of the world. Without plumbers, raw sewage would run rampant, and living conditions would deteriorate quickly. Apprenticeship programs exist to train people in the art of plumbing. Apprentices are taught much more than how to solder a joint or grade a pipe. They are taught math, science, and code requirements among other things. It takes more to be a good plumber than a strong back and a willingness to work. Today's plumbers have more rules and regulations to follow than ever before, and they must be capable of comprehending and abiding by the plumbing code.

Because of the complexities of modern plumbing, apprentices must dedicate hundreds of hours to classroom training. First-year students are likely to spend more than 150 hours in the classroom. What will these students study? The pupils will spend about 30 percent of the classroom time learning installation practices. Mathematics will account for about 20 percent of the teaching time. Another 20 percent of the time will be spent studying health and safety. Learning about related science, first-aid, and blueprints will consume the remainder of the schedule.

Some apprentices go to school during the day, and others attend classes at night. The night classes may be held only once a week and last between 2 and 3 hours. Apprenticeship programs are flexible enough to allow almost anyone the opportunity to learn new skills.

Many apprentices already have jobs in their chosen field. They work as apprentices or helpers during the day and study at night. These people have the benefit of hands-on training during their classroom learning. As you will discover in the next chapter, this is a valuable asset.

Plumbing apprentices are both male and female, and there can be a large range in the age of students. Many students are just getting out of high school, but there are some students that are looking for career changes. The students working to switch careers can be considerably older. Some apprentices are still in high school. They get in their classroom hours in an effort to enter the trade as soon as they graduate from high school. Plumbing apprentices come in all ages, but they all have the same goal—to become a licensed plumber.

Plumbers are frequently the subject of many monetary jokes. People like to joke about how all plumbers are rich and make more money than doctors and attorneys. While I don't believe this to be true, it is a fact that plumbers can make very good livings. In fact, plumbers are generally near the top of the list for highest paid trades. When a plumber progresses to the top level of the trade and becomes a master plumber, new opportunities abound. Master plumbers can go into business for themselves, and their income is potentially unlimited.

Money is certainly one reason for entering the plumbing trade, but it is not the only reason. Individuals who enjoy working with their hands will like plumbing; there is a lot of hands-on work in the trade. Plumbing is rarely boring. The job presents new challenges and opportunities with every new day, and owning your own plumbing business is a good goal to work toward. There are countless reasons why plumbing is a respectable and worthwhile career.

If you think plumbers spend all their time working in sewers and ditches, you are wrong. Most plumbers spend very little time in sewers and ditches. There are many different types of plumbing, and job opportunities abound in each of them. Some plumbers spend their days installing gold faucets that cost more than $2500. Many plumbers fill their work days with repair jobs on existing plumbing. Some plumbers find drain cleaning to be a very profitable way to spend the day. Plumbers who do new construction never handle a dirty toilet or a sewer machine. They spend their days installing new plumbing. There is a great deal of diversity in plumbing, and good apprentices can work toward specializing in the segment of their choice. It is, however, important for apprentices to learn all aspects of plumbing.

You will be taught many skills in the classroom. Most classes will be set up to give you some hands-on experience, and you will derive extensive knowledge about the plumbing trade. Let me give you a quick outline of the types of subjects you will cover in an average apprenticeship class:

Safety Safety is a top requirement in the plumbing trade. Being a plumber can be dangerous for you and the people around you. If you are not properly trained in the use of tools and equipment, fingers can be lost, and people can be killed. Normally, an apprentice is shown how to use the tools of the trade safely. First-aid techniques are generally part of the apprentice's learning experience, and other safety concerns are also covered.

Science Science may not be the first thought that comes to mind when you think of plumbing, but it is an element of the trade. Factors like gravity, vacuums, gases, and siphonage all come into play with plumbing. Apprentices will learn what's needed from related sciences.

Tools Tools are needed for nearly every phase of plumbing. Knowing which tool to use and how to use it properly is half the battle in plumbing. Apprentices receive extensive training in the selection and use of tools.

Blueprints If you will be doing new installations, you will need to be able to read blueprints. This is especially true when working with commercial plumbing installations. A skilled plumber can take a set of blueprints and plan an entire job without ever leaving the office. This is not to say that all the plans will work out on the job, but the basic plan of attack can be formed well before the job is started.

Having the capability to read blueprints is also needed to make take-offs and cost estimates. A take-off is a list of materials that will be needed to complete a job. Experienced plumbers can review a set of blueprints and develop the list of materials they will need. As an apprentice, you will learn what blueprint symbols mean and how to interpret what the drawing is showing you.

Math Math can play a big role in plumbing, and I don't mean just knowing how to read a tape measure. Before you are able to pass the test for a master's license, you will need to have a thorough knowledge of math. Now don't get me wrong, you don't have to be a wizard of calculations, but you will need to know more than how to add and subtract. You will have to learn to figure cubic measurements, how to find surface areas, and so forth. A good apprenticeship program will provide you with this needed knowledge.

Code requirements Plumbers must constantly abide by plumbing code requirements. If you pick up a code book and thumb through it, you will see that it is not casual reading. The plumbing codes are complex and at first often seem to be written in a foreign language. Your apprenticeship instructor will ease the burden of deciphering the cryptic text of the plumbing code.

Installation procedures Installation procedures will, of course, be a part of your classroom training. You should be taught about underground, rough-in, and finish plumbing. While you are studying these phases of the trade you are likely to learn about well, septic, and conditioning systems and much more. Gas piping is also likely to be a part of your installation instruction.

There is much to tell about on-the-job training, so much, in fact, that we will move to the next chapter to explore it.

Chapter

2

Field Experience

Field experience can never be replaced in a classroom. There is no substitute for real-life experience. It is easy to teach theory and to show field techniques in the classroom. However, there are some things that just can't be taught adequately during a school course. The difficulty in simulating some of the obstacles plumbers must face on a daily basis creates a need for in-field experience.

Even if you spend 4 years in a good apprenticeship program, you will not be prepared for the challenges you will meet on the job. You will be well prepared with the basics for facing the problems in the field, but until you get your hands on some real plumbing, you will not be ready to work on your own. The classroom instruction will prepare you for the job, but it will leave some voids that only on-the-job experience can fill in. You *can* go straight to work after graduation, but there will be many times when you will become confused and frustrated. Plumbing in the real world doesn't always work the way it does in the classroom. Let me give you some examples of what I'm talking about.

New Installations

The procedures for new installations will be covered in your apprenticeship class. You will learn how to join pipes and how much fall each pipe should have on it. There will be plenty of discussion on code requirements and job safety. But will the instructor prepare you for dealing with the other trades involved in the job you are working on? Will you be taught how to work with electricians and heating, venting, and air conditioning (HVAC) mechanics to avoid conflicts? If you are lucky, you will get a quick once-over on the ethics and principles of maintaining harmony between the trades, but you won't be prepared for the real thing.

Trade conflicts are common, and trade wars can break out. An inexperienced plumber can create animosity among the other trades very quickly. Not only can this be hazardous to your health, but it can make your job much harder. Until you are out there, working on jobs with other trades, you will not have the experience needed to make your installations go smoothly. Here are a few quick examples of how other trades can affect your plumbing installation.

Fairly early in my career I plumbed a house and had a terrible experience. I had completed all the rough-in plumbing and the job had been inspected and approved by the local plumbing inspector. After the inspection, I removed the test gear and went on to the next job. Several weeks later it was time to go back to that house and set the plumbing fixtures.

I went to the house and set all the fixtures. After checking all my connections, I turned the main water supply on. The main cut-off valve was in the basement of the home. Water rushed into the pipes, as it was supposed to, but something was wrong. The water continued to run for much too long. I thought perhaps I had left an outside hydrant open. After going outside and checking the two wall faucets, I found no evidence of a leak. I reentered the basement and could still hear water rushing through the pipes. Slightly panicked, I ran upstairs.

When I got to the first floor, the gushing sound of water was much louder; it was in the ceiling: between the first and second floor. I ran back to the basement and cut the water off. I couldn't understand what had happened; after all, the job had been tested and inspected.

After cutting into the brand new, and now soaked, ceiling, I discovered the problem. There were two ½-inch water pipes in the ceiling that had been blowing water all over the place. Now, keep in mind, this was in a new house, with finished carpeting on the floors, a finished paint job, and people anxious to move in. All of a sudden, I was in serious trouble.

How did my pipes spring leaks? Well, they didn't do it by themselves. After removing a significant section of the ceiling, it became obvious what had happened. The open ends of my pipes were rough and jagged. The pipes had been cut, probably with a chain saw. What kind of person would have vandalized the house in this way? It wasn't a vandal at all, it was an HVAC mechanic.

My pipes had been installed in a way that obstructed the joist bay for the HVAC mechanic. Instead of asking to have the pipes relocated, the sheetmetal worker slashed my pipes and removed the sections that had been in the way. No one ever told me of the need to repair the pipes. When the drywallers hung the ceiling, they didn't say anything about the mangled pipes. It was not until the house was finished and the fixtures set that I discovered the problem.

While this incident wasn't really my fault, I was caught in the middle of it. Everyone involved was upset over the damage and delays. Could this problem have been avoided? Yes, it could have. I could have gone over my piping diagram with the other trades to make sure it wouldn't create any conflicts. Did anyone ever teach me this lesson in school? No, I had to learn it the hard way.

Was this the only time I experienced problems with other trades? No. I've dealt with problems arising from carpentry, electrical, foundation, roofing, and most other trades. As I gained field experience, I learned how to minimize the conflicts, but they still happen, occasionally.

Electricians, carpenters, and HVAC mechanics are the ones plumbers have the most trouble with. Without cooperation, these trades can create nearly impossible working conditions for each other. Carpenters can put load-bearing timbers where the plumber has to cut a hole for toilet or tub waste. Plumbers can block paths for duct work and burn electrical wires with their torches. If I were to list every conceivable way for the various trades to mess each other up, I would need the space of this whole book to do it.

When you begin to work in the field, pay attention to the other trades. Look before you leap, and try to avoid problems by communicating with the other contractors. When you need to cut through a floor joist for your toilet rough-in, talk to the lead carpenter first. The carpenter may want to head off the joist before you sever it. Cooperation and communication are two of the first skills you must learn to do well in the field.

Repair Work

Repair work can be a big part of your plumbing career. Some plumbers specialize in repairs and do little else. You will, no doubt, learn about making repairs in the classroom, but you will not be prepared for what you will find on some jobs. Remember, what works in the classroom will not always work in the field. Let's check out some examples of how the conditions in the field can alter your conception of classroom plumbing.

Repair work can cover a broad spectrum of plumbing fixtures and devices. You might be working with bed-pan washers one day and automatic pig-watering devices the next. Classroom instruction is limited in what it can show you on repairs. For our first example, let's take the replacement of a flush valve under consideration.

When your instructor showed you how to replace a flush valve in the classroom, the job was quick and easy. You unbolted the toilet tank from the bowl, loosened the retaining nut on the flush valve, and in a matter of a few minutes, you had made the successful substitution of the old with the new.

Now you are out in the field. You are on your first service call to replace a flush valve. Being a recent graduate of apprenticeship school, you are not concerned about the task at hand. You get to the job and proceed with your work.

The first problem you run into is a cut-off valve that won't cut the water off; this never happened in school. You conquer the stubborn valve by cutting off the main water valve, and you are back at work. When you put your screwdriver on the tank-to-bowl bolts, the heads disintegrate into floating particles of rust. Hey, nobody ever prepared you for this to happen. After using your creative genius, you saw off the tank-to-bowl bolts and remove the toilet tank. Ah, now you're home free, right? Wrong; read on.

The flush valves you worked with in school were made of a nylon-type material. The plastic nuts turned easily, and you had no problem with them. But this toilet has a metal flush valve and retaining nut. The big brass nut hasn't been loosened in years, and it is on to stay.

After repeated attempts to turn the nut, you feel like giving up. You know that too much stress on the part will break the toilet tank, and you don't want that to happen. You've even tried putting a pipe wrench on the nut with no luck. Then you decide to squirt the threads with a spray that is supposed to loosen the nut. You wait the required time after spraying the part, but you still can't turn the nut. The homeowner is standing behind you and wondering if you know what you're doing.

Okay, enough of the suspense. Let me tell you what to do. Some plumbers would use their torches to heat the brass. The heat will burn out most of the crud holding the connection and allow the nut to be turned, but this is dangerous. Too much heat will cause the toilet tank to pop, sending china missiles flying toward your face. You will have a broken tank, an angry homeowner, and possibly some serious injuries.

You could put two wrenches on the flush valve one on the nut and the other on the body of the valve. This double-wrench system might allow enough leverage to make removing the nut possible.

If all else is failing, you can reach for your trusty hacksaw blade. It is hoped that you have a plumber's hacksaw, the type that allows multiple positions of the blade. The hacksaw blade will cut through the brass body of the flush valve without so much risk to the tank or your eyes.

The point to these notes on alternative removal methods is to stress that most classrooms are not equipped with seized flush valves. You can learn the standard procedures in class, and you might get some verbal instruction on complications, but until you face it in the field, you won't have the experience of dealing with a frustrating flush valve.

Other repairs that can quickly go astray include water heaters that won't drain, faucet nuts that have rusted to the sink, soldering pipes with water in them, and so on. There are plenty of circumstances in the field to test your mettle. Most of these problems can't be taught effectively in the classroom; you have to learn them on the job.

Learn from the Pros

As an apprentice, you are entitled and expected to learn from the pros. You are essentially working two jobs. Your first job is to do what your plumber tells you to do. In addition to that job, you must endeavor to learn what the plumber is doing and why it is being done; you will have to do this without getting in the way.

When I entered the plumbing trade, plumbers took the time to teach their helpers how and why a job was done. Over the years, production and the increased desire for making more money has reduced the time most plumbers are willing to devote to apprentices. Time is money, and the time a plumber spends training an apprentice is likely to result in lost income.

In the old days, a helper was trained to be a plumber, and then the trainee stayed with the employer and worked as a plumber. A lot has changed in the last couple of decades. Trainees are not as loyal as they once were. Many of today's trainees will change jobs to make a few extra bucks a week.

When you consider the time and money a plumber spends training an apprentice, you can see why the plumber might be hesitant to invest much when it is anticipated that the apprentice will change jobs at the drop of a hat. The combination of trainees jumping from job to job and plumbers being reluctant to invest in unknown personnel has created a serious problem. The amount and quality of on-the-job training are suffering from the effects of a lack of loyalty.

When plumbers won't train apprentices, the apprentices have their work cut out for them. There is going to be a lot of learning from hard knocks and faulty plumbing. The reputation of the trade will suffer, and the number of qualified plumbers will dwindle. On the whole, this is bad news. However, for the apprentices who become good plumbers, the news is not all bad. Good plumbers are always in demand. The best plumbers can demand, and get, high wages. If you spend the time to pay your dues early, you can reap the rewards later.

I'm not saying plumbers shouldn't spend time and money training new plumbers. I wish modern apprentices had the opportunities to learn from skilled masters as I did. In some ways, modern apprentices have it better. They have the advantage of classroom training that I never got. Still, the lack of field training is serious business.

I understand the plight of the plumbing contractor. As a business owner, I know the costs involved with training. As a teacher of plumbing courses, I've seen the lack of field training being given to today's trainees. My experiences on both sides of the table shows me the cause and effect of what is happening.

When you go into the field for on-the-job training, you are going to have to want to learn. It will be up to you to take the initiative and carve out your field education. If all you do is stand around, waiting for quitting time, you are not going to move up the plumbing ladder very quickly. On the other hand, if you are motivated and show promise, your lead mechanic will probably help you. The choice is yours, but if you don't put out the extra effort to learn, you can't expect the plumber to spend the extra money and time to train you.

What is the best way to learn from the pros?

There is no best way to learn from the pros. Much of your learning process in the field will depend on the type of work you are doing and the person responsible for training you. The other variable will be you. In general, you should be able to learn the most by paying attention to what is being done and asking questions when you don't understand a procedure.

Most seasoned plumbers have plenty of old war stories they enjoy telling, and you can learn from these sage words. Strike up a conversation while riding to the job, taking breaks, and any other time when talking is appropriate. The plumbers will probably talk your ear off but pay attention, and you will likely learn a lot. You may have to sift through some of the stories to find the truth, but most of them will be served up with a heavy dose of experience—experience you need. Let me give you a couple of ideas of what these stories might be about.

If I were to sit in the truck with you at lunch, I could tell you stories every day, for at least a year, without repeating myself. I could tell you about the unfortunate helper who removed a cleanout plug the wrong way and got a face full of raw sewage. Then there's the story about a helper who pulled a test plug and was trapped in an outside foundation hole filled with three-stories worth of test water on a day with freezing temperatures. You might learn how one of my helpers was cutting a vent hole in a roof and let the drill knock him out of the attic for a long fall to the floor below. I could make hand gestures to illustrate how countless helpers have gotten sore jaws when their right-angle drills hit knots and nails in the wood they were drilling.

After catching my breath, I could tell you about the time I let a big drill bit get caught in my shirt and how it nearly paralyzed my right arm. While on the subject of personal injuries, I could tell you about how a reckless moment with the drill separated my stomach muscles and caused me great pain for 6 months.

Then there was the time the rope I had tied around the chimney pulled the old bricks loose and sent me sliding off the tin roof. I could tell you in great detail how it feels to lose a whole load of 6-inch pipe off your truck in the middle of rush-hour traffic. Oh, the stories I could tell. Anyway, you get the point. Experienced plumbers can share a wealth of knowledge with you in short stories. If you listen carefully, you can learn from their mistakes and avoid many of the problems they have encountered.

Trial by fire

Most plumbers, for whatever reason, can't resist giving the rookie a trial by fire. As the new kid on the block, you will bear the brunt of job-site pranks. These lessons are usually harmless, but they help you learn.

Trial and error

There will be plenty of times when you will learn by trial and error. Seasoned plumbers know that the lessons that stick with you are the ones that are learned the hard way. Most plumbers will allow apprentices to make mistakes to learn. But they should protect you from dangerous experiences, like cutting a cast-iron stack the wrong way. As you struggle to overcome your own job-site problem, the plumber will probably be amused. Maintain your composure and earn the plumber's respect. This type of learning experience is not a childish prank; it is effective training.

Go with the flow

When it all washes out, you will have to go with the flow. All plumbers train their apprentices a little differently. In most cases you can expect the hardest and least desirable work available for a while. There seems to be an unwritten code among plumbers that dictates that apprentices must live a hard life in the beginning. Survive this initial test of your strength and attitude, and you will be on your way to becoming a great plumber. Fail this test, and you will be looking for another company to work for.

A big part of your field training will involve working with the plumbing code. The code will affect most of your daily activities. As an apprentice, you will learn about the plumbing code in the classroom and in the field. To get a preview of how the code will affect your career, turn to the next chapter.

Chapter

3

An Introduction to the Plumbing Code

This chapter is an introduction to the plumbing code. There are actually several plumbing codes. Various regions of the country adapt one of the codes for use in their jurisdictions. The boilerplate code may be amended by the local authorities to create modifications as necessary. Since every jurisdiction can have variations in its plumbing code, it is difficult to predict exactly what will be required as you move from jurisdiction to jurisdiction. Even so, there are many similarities to the three major codes. Once you learn the basics, a quick trip to the local plumbing inspector's office can answer your questions on the particulars of any existing modifications.

This chapter is going to deal with the basics. Before you take these words for gospel, check your local code requirements. I have a strong understanding of all the major plumbing codes, but I cannot predict the various changes your local authorities may have substituted in their plumbing code.

The Code Book

Every apprentice should have a copy of the code book. Even though you may not have a clue about what is being said in the text, you need the book. A phone call to your local code enforcement office will tell you where and how to obtain a copy of the plumbing code.

The code book should be carried with you to class and out on the job. When your boss tells you to install the hangers on the drain pipes being installed, you can refer to the book for the required spacing of the hangers. You could ask the plumber, but if you already know how to space the hangers, you will move up a step in the plumber's esteem. It would be a good idea to confirm your interpretation of the code with the plumber, but it will show well for you if you know the right answer.

Code books can be hard to understand, even for experienced plumbers. There are other books available to help you understand what you are reading in the code book. I wrote one of these books, but there are also others available. If you would like my book, *The Plumbing Codes Handbook,* on interpreting the plumbing codes, you can order it from McGraw-Hill, Inc.

Once you have your code book, you can begin studying. The more studying you do on your own, the better off you will be. Jot down questions about the code as you run into them. When you get the opportunity in class or on the job, ask for clarification of your questions. Self-study is a very effective way to move to the head of the class, both in the classroom and on the job.

When you begin reading your code book, you will run into plenty of questions. Don't be intimidated. The code is complex, but most of it is logical. Once you learn the basics of the trade, the code will make a lot more sense. After you are licensed and in the field, you will only use a small portion of the code. But until you are licensed, you must learn all of it or at least enough of it to pass your exam. Let me give you a broad idea of what you will find in a code book.

Administration

Administration is one of the first elements of the plumbing code that you will encounter. All the major plumbing codes include sections on administration that include many aspects of the plumbing code. Let me fill you in on what some of them are.

Permits Permits are covered when you read about administration. You will learn who can obtain permits and what jobs require the use of permits.

Tests The tests you read about in the administration section of the code book are not the exams you take for licensing. These tests are what is required for testing your plumbing installations.

Appeals As you read your code book, you will find a section on appeals. This part of the code book tells how you can address disagreements in the interpretation of the code.

Violations The text that talks about violations will inform you of what a violation of the code is. All of the code book is important reading, but identifying violations is particularly important.

Penalties If you fail to recognize and avoid code violations, you can read about the penalties in this section of the administrative chapter.

Blueprints and specifications Blueprints and specifications are often required for new installations. Drawing riser diagrams and submitting a plan for the work you will be doing is generally necessary for obtaining plumbing permits. This section in the administration chapter of a code book will tell you what types of drawings and specifications are required for a permit application. You will learn how critical the drawings and specifications are to your career.

Inspections The requirements for inspections conducted by code officers are covered in the inspection segment of the administration chapter. You will learn what inspections are required and when they are required.

Fees The fees charged for permits and similar needs will be explained in this section of your administration chapter.

Existing conditions Existing conditions can alter the way you proceed with a job. Many times existing plumbing is not in compliance with current code requirements. Normally, if the existing plumbing is not hazardous, it may be left in its present condition. The segment of the administration chapter dealing with existing conditions will clear up the confusion many plumbers have over existing conditions.

Definitions

Definitions are a common element of the various plumbing codes. There is not a lot to say about this section of the code books. The definitions are just that: definitions.

General information

When you look under general information in the code book, you will find a number of topics. Some of the highlights of the general information will deal with the following: health, safety, sleeves, pipe protection, fittings, trenches, and so on. There are plenty of topics to keep you busy in the general information section of the code book.

Specific chapters

There are many specific chapters in the various code books. You will find chapters that delve deeply into all aspects of the plumbing trade. These chapters detail everything from materials to hospital plumbing.

You will learn about drainage systems, potable water systems, fixture traps, vents, fixtures, inspections, and engineered plumbing systems.

Charts and tables

When you thumb through your local code book, you will see numerous charts and tables. These illustrations are not there simply to break up the text; they are an integral part of the plumbing code. Many of the tables will give you specific code requirements. Some of the illustrations will assist you in putting the code into practice. You cannot skip the illustrations when you are mastering the plumbing code.

What else is in a code book?

You will find a host of useful information in your code book. There will be hundreds of pages of rules, laws, and regulations to digest. Not all jurisdictions have plumbing laws in the code; instead their requirements are rules and regulations. If you violate them, you can lose your license, but you will not have committed a crime. Other areas incorporate laws into their plumbing codes. The difference between a rule and a law is substantial, but in the case of the plumbing code, they are both words to work by.

The most important aspect of your code book is your career. The book holds the answers to your questions and provides the proper path for you to walk. Buy a code book and study it. If you plan to be a professional plumber, you should commit the code to memory. If your memory is not too good, never allow your code book to be far away.

How Important Is the Plumbing Code?

The plumbing code is essential in establishing and maintaining a quality environment. Without the plumbing code, our lives would be considerably different. For plumbers, the code is a way of life.

The code is not just a bunch of rules and regulations to run up the cost of plumbing jobs; it deals with issues that could very quickly turn into serious problems. Failure to comply with the plumbing code can result in serious injury or death. The code is definitely not just a group of rules made to make life difficult for plumbers.

There are many subtle ways to protect the health and lives of the public with the plumbing code. Some code issues capable of preventing serious accidents are easy to identify. Many plumbers don't think much about the importance of adhering to the code. This group sees the code as an inconvenience. If they can bypass the code or slip through an inspection, they don't feel any concern or remorse. This group of plumbers is dangerous. Let me tell you why.

Water heaters

Water heaters are involved in many code violations. The most common violation involving water heaters is the failure to obtain permits to install them. Permits are required for all water heater installations. It doesn't matter whether you are replacing an old water heater or installing new plumbing; a permit is required. The installation of the water heater is required to be inspected and approved by a code enforcement officer. Even though the code is quite clear on this issue, a lot of plumbers never bother with permits for replacement water heaters.

Why wouldn't licensed plumbers get permits for replacement water heaters? The answer to this question is time and money. A good plumber can swap out a water heater in less than 2 hours. If plumbers don't get permits for this type of work, they are in and out in less than 2 hours, and they can keep their prices down and their profits up.

When plumbers go by the book on water heater replacements, their fees must go up. This can mean lost jobs in a competitive market where not all plumbers are playing by the rules. Homeowners don't know permits are required; they look only at the amount of money they will have to spend for a new water heater. Obviously, the companies cheating the system can sell their jobs for less. They don't have to pay the price of the permit, and they don't have to lose time in obtaining the permit and the inspection.

With money being the major motivation of most businesses, it stands to reason that if the competition is cutting corners and bootlegging jobs, some other companies are going to follow suit. This snowball effect creates a situation where it is almost common practice to install replacement water heaters without permits. Not only is this in violation of the plumbing code, it is dangerous.

When licensed plumbers violate the code, they are putting their licenses at risk. If they have too many serious violations, they could lose their licenses and their livelihood. You can compare it to drunk driving. When people drive drunk, they are risking lives and their driving licenses. Plumbers that install water heaters without permits are doing the same type of thing.

Paying fines and losing licenses are the small penalties for bootlegging the installation of water heaters. What happens if the installation results in a death? Not only will the offending plumber have to live with the fact that his or her work took someone's life, he or she will probably be ruined financially. When plumbers are sued, their insurance is not likely to cover their losses because they didn't follow proper procedures in the installation. The fact that the installation was in violation of the code and a matter of gross negligence by the plumber is all the excuse the insurance company will need to wash their hands of the claim.

How could a simple water heater installation kill someone? Water heaters are required to be equipped with temperature and pressure relief valves. If these valves are not installed or if the valves are not of the proper rating, the water heater can become a bomb. Without the right relief valve, a water heater can explode. The power from such an explosion can be massive. It can devastate a home and anyone who happens to be in the vicinity. These explosions do happen, and their destruction has been documented in many cases.

It is possible for any water heater to blow up. Even when a relief valve of the proper rating is installed, there is a remote chance something will malfunction and cause a problem. So, even if bootlegging plumbers install water heaters properly, they may still be at risk of more than fines and lost licenses.

When plumbers make legal installations and have their work inspected, they greatly reduce their exposure. If a good installation results in a serious problem, the plumbers can document their actions. They can call upon the code enforcement office for documentation of their proper installations and inspections. If they are not exonerated, at least their insurance companies will have to pay for the damages.

So, you see, something as simple as replacing a water heater can become very serious. The code is full of regulations that can improve and save lives. When you become a professional plumber, you should appreciate what the code and the plumbing inspectors do for you.

Backflow prevention

Backflow prevention is another section of the plumbing code that can have a major impact on public health. The installation requirements for backflow prevention devices are in the code to limit the contamination of potable water systems. Without the proper protection, a large portion of a city's water system could be polluted from a single incident. Let me give you an example.

When a large water main breaks, a vacuum can be formed, sucking water back into the system from sources it has supplied. In other words, if the municipal water main in the street were to break, the water in the pipes of houses served by the main could be pulled back into the main. If the water in the house pipes was contaminated, the contamination could spread into the main water system. This could be particularly bad if the water main was feeding a company that worked with strong chemicals.

Can you imagine the effect of having acid sucked back into the city water system? Well, without backflow preventers, it could happen. There are many ways and circumstances where plumbing that is not protected from siphonage can cause serious trouble.

Backflow preventers and water heaters are only two issues in the code that could affect lives; there are many others. I picked these two for examples because they are the two that are frequently ignored. Please, don't assume the code is just a bunch of mumbo-jumbo. It is a very important element in maintaining a good quality of life.

Now that we have outlined the plumbing code, let's move on to Chapter 4 and get on with specific issues that will help you become a professional plumber.

Chapter 4

Safety

Safety should be a consideration in everything we do. The old saying, "An ounce of prevention is worth a pound of cure," is well worth remembering. Safety options are all around us, and yet, millions of people suffer each year because they failed to take advantages of the resources available to them. Most pain and death from accidents is preventable. How it is prevented depends on the context of the danger. This chapter is going to give you advice on how to enjoy a safe career as a plumber. Some of the points will be pure, old-fashioned common sense, and some will be more complex. In all cases, the advice is meant to spare you, and others, the agony of injury and the loss of income or life that often comes with serious accidents.

Plumbing can be a very dangerous trade. The tools of the trade have the potential to be killers. The requirements of the job can place you in positions where a lack of concentration could result in serious injury or death. The fact that plumbing can be dangerous is no reason to rule out the trade as your profession. Driving can be extremely dangerous, but few people stay out of automobiles out of fear.

Fear is generally a result of ignorance. When you have a depth of knowledge and skill, fear begins to subside. As you become more accomplished at what you do, fear is forgotten. While it is advisable to learn to work without fear, you should never work without respect. There is a huge difference between fear and respect.

If, as an apprentice, you are afraid to climb up on a roof to flash a pipe, you are not going to last long in the plumbing trade. However, if you scurry up the roof recklessly, you could be injured severely or perhaps, even killed. You must respect the position you are putting yourself in. If you are using a ladder to get on the roof, you must respect the outcome that a mistake could have. Once you are on the roof, you must be conscious of footing conditions and the way that you negotiate the pitch of the roof. If you pay attention, are properly trained, and don't get careless, you are not likely to get hurt.

Being afraid of the roof will limit or eliminate your plumbing career. Treating your trip to and from the roof like a walk into your living room could be deadly. Respect is the key. If you respect the consequences of your actions, you are aware of what you are doing and the odds for a safe trip improve.

Many young apprentices are fearless in the beginning. They think nothing of darting around on a roof or jumping down in a sewer trench. As their careers progress, they usually hear of or see on-the-job accidents. Someone gets buried in the cave-in of a trench. Somebody falls off a roof. A metal ladder being set up hits a power line. A careless plumber steps into a flooded basement and is electrocuted because of submerged equipment. The list of possible job-related injuries is a long one.

Millions of people are hurt every year in job-related accidents. Most of these people were not following solid safety procedures. Sure, some of them were victims of unavoidable accidents, but most were hurt by their own hands, in one way or another. You don't have to be one of these statistics.

In nearly 20 years of plumbing, I have only been hurt seriously on the job twice. Both times were my fault. I got careless. In one of the instances, I let loose clothing and a powerful drill work together to chew up my arm. In the other incident, I tried to save myself the trouble of repositioning my stepladder while drilling holes in floor joists. My desire to save a minute cost me torn stomach muscles and months of pain from a twisting drill.

My accidents were not mistakes; they were stupidity. Mistakes are made through ignorance. I wasn't ignorant of what could happen to me. I know the risk I was taking, and I knew the proper way to perform my job. Even with my knowledge, I slipped up and got hurt. Luckily, both of my injuries healed, and I didn't pay a lifelong price for my stupidity.

During my long plumbing career I have seen a lot of people get hurt. Most of these people have been helpers and apprentices. Of all the on-the-job accidents I have witnessed, every one of them could have been avoided. Many of the incidents were not extremely serious but a few were.

As an apprentice, you will be doing some of the most dangerous work in the trade. You will be drilling holes, running threading machines, snaking drains, installing roof flashings, and a lot of other potentially dangerous work. It is hoped that your employer will provide you with quality tools and equipment. If you have the right tool for the job, you are off to a good start in staying safe.

Safety training is another factor you should seek from your employer. Some plumbers fail to tell their apprentices how to do jobs safely. It is easy for someone, like an experienced plumber, who knows a job inside and out, to forget to inform an inexperienced person of potential danger.

For example, a plumber might tell you to break up the concrete around a pipe to allow the installation of a closet flange and never consider telling you to wear safety glasses. The plumber will assume you know the concrete is going to fly up in your face as it is chiseled up. However, as a rookie, you might not know about the reaction concrete has when hit with a cold chisel. One swing of the hammer could cause extreme damage to your eyesight.

Simple jobs, like the one in the example, are all it takes to ruin a career. You might be really on your toes when asked to scoot across an I-beam, but how much thought are you going to give to tightening the bolts on a toilet? The risk of falling off the I-beam is obvious. Having chips of china, from a broken toilet on which the nuts were turned one time too many, flying into your eyes is not so obvious. Either way, you can have a work-stopping injury.

Safety is a serious issue. Some job sites are very strict in the safety requirements maintained. But a lot of jobs have no written rules of safety. If you are working on a commercial job, supervisors are likely to make sure you abide by the rules of the Occupational Safety and Health Administration (OSHA). Failure to comply with OSHA regulations can result in stiff financial penalties. However, if you are working in residential plumbing, you may never set foot on a job where OSHA regulations are observed.

In all cases, you are responsible for your own safety. Your employer and OSHA can help you to remain safe, but in the end, it is up to you. You are the one who has to know what to do and how to do it. And not only do you have to take responsibility for your own actions, but you also have to watch out for the actions of others. It is not unlikely that you could be injured by someone else's carelessness. Now that you have had the primer course, let's get down to the specifics of job-related safety.

As we move into specifics, you will find the suggestions in this chapter broken down into various categories. Each category will deal with specific safety issues related to the category. For example, in the section on tool safety, you will learn procedures for working safely with tools. As you move from section to section, you may notice some overlapping of safety tips. For example, in the section on general safety, you will see that it is wise to work without wearing jewelry. Then you will find jewelry mentioned again in the tool section. The duplication is done to pinpoint definite safety risks and procedures. We will start with general safety.

General Safety

General safety covers a lot of territory. It starts from the time you get into the company vehicle and carries you right through to the end of the day. Much of the general safety recommendations involve the use of common sense. Now, let's get started.

Vehicles

Many plumbers are given company trucks for their use in getting to and from jobs. As an apprentice, you are not likely to have your own truck, but you will probably spend a lot of time loading and unloading company trucks. And, of course, you will spend time either riding in or driving them. All of these areas can threaten your safety.

If you will be driving the truck, take the time to get used to how it handles. Loaded plumbing trucks don't drive like the family car. Remember to check the vehicle's fluids, tires, lights, and related equipment. Many plumbing trucks are old and have seen better days. Failure to check the vehicle's equipment could result in unwanted headaches. Also, remember to use the seat belts; they do save lives.

Apprentices are normally charged with the duty of unloading the truck at the job site. There are a lot of ways to get hurt in doing this job. Many plumbing trucks use roof racks to haul pipe and ladders. When you unload these items, make sure they will not come into contact with low-hanging electrical wires. Copper pipe and aluminum ladders make very good electrical conductors, and they will carry the power surge through you on the way to the ground.

If you are unloading heavy items, don't put your body in awkward positions. Learn the proper ways for lifting, and never lift objects inappropriately.

If the weather is wet, be careful climbing on the truck. Step bumpers get slippery, and a fall can impale you on an object or bang up your knee.

When it is time to load the truck, observe the same safety precautions you did in unloading. In addition to these considerations, always make sure your load is packed evenly and well secured. Be especially careful of any load you attach to the roof rack, and always double check the cargo doors on trucks with utility bodies.

It will not only be embarrassing to lose your load going down the road, it could be deadly. I have seen a one-piece fiberglass tub-shower unit fly out of the back of a pick-up truck as the truck was rolling up an interstate highway. As a young helper, I lost a load of pipe in the middle of a busy intersection. In that same year, the cargo doors on the utility body of my truck flew open as I came off a ramp onto a major highway. Tools scattered across two lanes of traffic. These types of accidents don't have to happen. As an apprentice, it is your job to make sure they don't.

Clothing

Clothing is responsible for a lot of on-the-job injuries. Sometimes it is the lack of clothing that causes the accidents, and there are many times when too much clothing creates the problem. Generally, it is wise not to wear loose-fitting clothes. Shirt tails should be tucked in, and short-sleeve shirts are safer than long-sleeved shirts when operating some types of equipment.

Caps can save you from minor inconveniences, like getting glue in your hair, and hard hats provide some protection from potentially damaging accidents, like having a steel fitting dropped on your head. If you have long hair, keep it up and under a hat.

Good footwear is essential in the trade. Normally a strong pair of hunting-style boots will be best. The thick soles provide some protection from nails and other sharp objects you may step on. Boots with steel toes can make a big difference in your physical well-being. If you are going to be climbing, wear foot gear with a flexible sole that grips well.

Gloves can keep your hands warm and clean, but they can also contribute to serious accidents. Wear gloves sparingly, depending upon the job you are doing.

Jewelry

On the whole, jewelry should not be worn in the workplace. Rings can inflict deep cuts in your fingers. They can also work with machinery to amputate fingers. Chains and bracelets are equally dangerous, probably more so.

Eye and ear protection

Eye and ear protection is often overlooked. An inexpensive pair of safety glasses can prevent you from spending the rest of your life blind (Fig. 4.1). Ear protection reduces the effect of loud noises. You may not notice much benefit now, but in later years, you will be glad you wore

Figure 4.1 Safety glasses. (*Model, Andrew Wallace*)

it. If you don't want to lose your hearing, wear ear protection when subjected to loud noises, like jackhammers and heavy equipment.

Knee pads

Knee pads not only make a plumber's job more comfortable, but they help to protect the knees. Some plumbers spend a lot of time on their knees, and pads should be worn to ensure they can continue to work for many years.

The embarrassment factor

The embarrassment factor plays a significant role in job-related injuries. People, especially young people, feel the need to fit in and to make a name for themselves. Plumbing is sort of a macho trade. It is no secret that plumbers often fancy themselves as strong human specimens. Many plumbers are strong. The work can be hard and in doing it, becoming strong is a side benefit. But you can't allow safety to be pushed aside for the purpose of making a macho statement.

All too many people believe that working without safety glasses, ear protection, and so forth makes them tough. That's just not true; it may make them appear dumb, and it may get them hurt, but it does not make them look tough. If anything, it makes them look stupid or inexperienced.

Don't fall into the trap so many young plumbers do. Never let people goad you into bad safety practices. Some plumbers are going to laugh at your knee pads. Let them laugh; you will still have good knees when they are hobbling around on canes. I'm dead serious about this issue. There is nothing sissy about safety. Wear your gear in confidence, and don't let the few jokesters get to you.

Tool Safety

Tool safety is a big issue in plumbing. Anyone in the plumbing trade will work with numerous tools. All of these tools are potentially dangerous, but some of them are especially hazardous. This section is broken down by the various tools used on the job. You cannot afford to start working without the basics in tool safety. Your apprenticeship instructor and your employer should cover this topic thoroughly. But they may not, and I may know something they won't think to tell you. The more you can absorb on tool safety, the better off you will be.

The best starting point is to read all literature available from the manufacturers of your tools. The people who make the tools provide some good safety suggestions with them. Read and follow the manufacturers' recommendations.

The next step in working safely with your tools is to ask questions. If you don't understand how a tool operates, ask someone to explain it to you. Don't experiment on your own; the price you pay could be much too high.

Common sense is irreplaceable in the safe operation of tools. If you see an electrical cord with cut insulation, you should have enough common sense to avoid using it. There are many other such examples that you will learn in the following text. In addition to this type of simple observation, you will learn some interesting facts about tool safety. Now, let me tell you what I've learned about tool safety over the years.

There are some basic principles to apply to all of your work with tools. We will start with these basics, and then we will move on to specific tools. Here are the basics:

- Keep body parts away from moving parts.
- Don't work with poor lighting conditions.
- Be careful of wet areas when working with electrical tools.
- If special clothing is recommended for working with your tools, wear it.
- Use tools only for their intended purposes.
- Get to know your tools well.
- Keep your tools in good condition.

Now, let's take a close look at the tools you are likely to use. Plumbers use a wide variety of hand and electrical tools. They also use specialty tools. So let's see how you can use all these tools without injury.

Torches

Torches are often used by plumbers on a daily basis. Some plumbers use propane torches, and others use acetylene torches. In either case, the containers of fuel for these torches can be very dangerous. The flames produced from torches can also do a lot of damage.

When working with torches and tanks of fuel, you must be very careful. Don't allow your torch equipment to fall on a hard surface; the valves may break. Check all your connections closely; a leak allowing fuel to fill an area could result in an explosion when you light the torch.

Always pay attention to where your flame is pointed. Carelessness with the flame could start unwanted fires. This is especially true when working near insulation and other flammable substances. Since moisture is retained in concrete, if the flame is directed at concrete, the intense heat can cause the moisture to force the concrete to explode.

It's not done often enough, but you should always have a fire extinguisher close by when working with a torch. If you have to work close to flammable substances, use a heat shield on the torch. When your flame is close to wood or insulation, try to remove the insulation or wet the flammable substance before applying the flame. When you are done using the torch, make sure the fuel tank is turned off and the hose is drained of all fuel. Use a striker to light your torch. The use of a match or cigarette lighter puts your hand too close to the source of the flame.

Welding equipment

Welding equipment requires many of the same safety procedures used with torches. But there are a few extra points to ponder. Welding tanks should be kept vertical and protected from falling and from heat. The use of pure oxygen adds a new dimension to the possible dangers. Pure oxygen can cause spontaneous combustion when combined with some materials.

Lead pots and ladles

Lead pots and ladles offer their own style of potential danger. Plumbers today don't use molten lead as much as they used to, but hot lead is still used. It is used primarily to make joints with cast-iron pipe and fittings.

When working with a small quantity of lead, many plumbers heat the lead in a special ladle. They melt the lead with a torch and pour it straight from the ladle. When larger quantities of lead are needed, lead pots are used. The pots are filled with lead and set over a flame. All of this type of work is dangerous.

Never put wet materials in hot lead. If the ladle is wet or cold when it is dipped into the pot of molten lead, the hot lead can explode. Don't add wet lead to the melting pot if it contains molten lead. Before you pour the hot lead in a waiting joint, make sure the joint is not wet. As another precaution, don't leave a working lead pot where rain dripping off a roof can fall into it.

Obviously, molten lead is hot, so you shouldn't touch it. Be very careful to avoid tipping the pot of hot lead over. I remember one accident when a pot of hot lead was knocked over on a plumber's foot. Let me just say the scene was terrifying.

Drills and bits

Drills have been my worst enemy in plumbing. The two serious injuries I have received were both related to my work with a right-angle drill. The drills used most by plumbers are not the little pistol-grip, hand-held types of drills most people think of. The day-to-day drilling done by plumbers and apprentices involves the use of large, powerful right-angle drills. These drills have enormous power when they get in a bind. Hitting a nail or a knot in the wood being drilled can do a lot of damage. You can break fingers, lose teeth, suffer head injuries, and a lot more.

As with all electrical tools, you should always check the electrical cord before using your drill. If the cord is not in good shape, don't use the drill.

Always know what you are drilling into. If you are doing new-construction work, it is fairly easy to look before you drill. However, drilling in a remodeling job can be much more difficult. You cannot always see what you are getting into. If you are unfortunate enough to drill into a hot wire, you can get a considerable electrical shock.

The bits you use in a drill are part of the safe operation of the tool. If your drill bits are dull, sharpen them. Dull bits are much more dangerous than sharp ones.

When you are using a standard plumber's bit to drill through thin wood, like plywood, be careful. Once the worm driver of the bit penetrates the plywood fully, the big teeth on the bit can bite and jump, causing you to lose control of the drill.

If you will be drilling metal, be aware that the metal shavings will be sharp and hot.

Power saws

Plumbers don't use power saws as much as carpenters, but they do use them. The most common type of power saw used by plumbers is the reciprocating saw. These saws are used to cut pipe, plywood, floor joists, and a whole lot more. In addition to reciprocating saws, plumbers use circular and chop saws. All of the saws have the potential to cause serious injury.

Reciprocating saws are reasonably safe. Most models are insulated to help avoid electrical shocks if a hot wire is cut. The blade is typically a safe distance from the user, and the saws are pretty easy to hold and control. However, the brittle blades do break. This could result in an eye injury.

Circular saws are used by plumbers occasionally. The blades on these saws can bind and cause the saws to kick back. Chop saws are sometimes used to cut pipe. If you keep your body parts out of the way and wear eye protection, chop saws are not unusually dangerous.

Hand-held drain cleaners

Hand-held drain cleaners don't get a lot of use from plumbers who do new-construction work, but they are a frequently used tool of service plumbers. Most of these drain cleaners resemble, to some extent, a straight hand-held drill. There are models that sit on stands, but most small snakes are hand-held.

These small-diameter snakes are not nearly as dangerous as their big brothers, but they do deserve respect. These units carry all the normal hazards of an electric tool, but there is more.

The cables used for small drain cleaning jobs are usually very flexible. The cables are basically springs. The heads attached to the cables take on different shapes and looks. When you look at these thin cables, they don't look dangerous, but they can be.

When the cables are being fed down a drain and turning, they can hit hard stoppages and go out of control. The cable can twist and kink. If your finger, hand, or arm is caught in the cable, injury can be the result. To avoid this, don't allow excessive cable to exist between the drain pipe and the machine.

Large sewer machines

Large sewer machines are much more dangerous than small ones. These machines have tremendous power, and their cables are capable of removing fingers. Broken bones, severe cuts, and assorted other injuries are also possible with big snakes.

One of the most common conflicts with large sewer machines is with the cables. When the cutting heads hit roots or similar hard items in the pipe, the cable goes wild. The twisting, thrashing cable can do a lot of damage. Again, limiting excess cable is one of the best protections possible. Special sleeves are also available to contain unruly cables.

Most big machines can be operated by one person, but it is wise to have someone standing by in case help is needed. Loose clothing results in many drain machine accidents. The use of special mitts can help reduce the risk of hand injuries.

Electrical shocks are also possible when doing drain cleaning.

Power pipe threaders

Power pipe threaders are very nice to have if you are doing much work with threaded pipe. However, these threading machines can grind body parts as well as they thread pipe. Electric threaders are very dangerous in the hands of untrained people.

It is critical to keep fingers and clothing away from the power mechanisms. The metal shavings produced by pipe threaders can be very sharp, and burrs left on the threaded pipe can slash your skin. The cutting oil used to keep the dies from getting too hot can make the floor around the machine slippery.

Air-powered tools

Air-powered tools are not used often by plumbers. Jackhammers are probably the most used air-powered tools for plumbers. When using tools with air hoses, check all connections carefully. If you experience a blow-out, the hose can spiral wildly out of control.

Powder-actuated tools

Powder-actuated tools are used by plumbers to secure objects to hard surfaces, like concrete. If the user is properly trained, these tools are not too dangerous. However, good training and eye and ear protection are all necessary. Misfires and chipping hard surfaces are the most common problem with these tools.

Ladders

Ladders are used frequently by plumbers, both stepladders and extension ladders. Many ladder accidents are possible. You must always be aware of what is around you when handling a ladder. If you brush against a live electrical wire with a ladder you are carrying, your life could be over. Ladders often fall over when the people using them are not careful. Reaching too far from a ladder can be all it takes to have a fall.

When you set up a ladder, make sure you do so properly. The ladder should be on firm footing, and all safety braces and clamps should be in place. When using an extension ladder, many plumbers use a rope to tie rungs together where the sections overlap. The rope provides an extra guard against the ladder's safety clamps failing and the ladder collapsing.

When using an extension ladder, be sure to secure both the base and the top. I had an unusual accident on a ladder that I would like to share with you.

I was on a tall extension ladder, working near the top of a commercial building. I was working with gas piping that penetrated the side of the building, near the roof. The top of my ladder was resting on the edge of the flat roof. There was metal flashing surrounding the edge of the roof, and the top of the ladder was leaning against the flashing.

There was a picket fence behind me and electrical wires entering the building to my right. The entrance wires were a good distance away, so I was in no immediate danger. As I worked on the ladder, a huge gust of wind blew around the building. I don't know where it came from; it hadn't been very windy when I went up the ladder.

The wind hit me and pushed me and the ladder sideways. The top of the ladder slid easily along the metal flashing, and I could not grab anything to stop me. I knew the ladder was going to go down, and I didn't have much time to make a decision. If I pushed off of the ladder, I would probably be impaled on the fence. If I rode the ladder down, it might hit the electrical wires and fry me. I waited until the last minute and jumped off of the ladder.

I landed on the wet ground with a thud, but I missed the fence. The ladder hit the wires and sparks flew. Fortunately, I wasn't hurt and electricians were available to take care of the electrical problem.

This was a case where I wasn't really negligent, but I could have been killed. If I had secured the top of the ladder, my accident wouldn't have happened.

Screwdrivers and chisels

Eye injuries and puncture wounds are common when working with screwdrivers and chisels. When the tools are used properly and safety glasses are worn, few accidents occur.

The key to avoiding injury with most hand tools is simply to use the right tool for the job. If you use a wrench as a hammer or a screwdriver as a chisel, you are asking for trouble.

There are, of course, other types of tools and safety hazards found in the plumbing trade. However, this list covers the ones that result in the most injuries. In all cases, observe proper safety procedures and use safety gear, such as eye and ear protection.

Coworker Safety

Coworker safety is the last segment of this chapter. I am including it because workers are frequently injured by the actions of coworkers. This section is meant to protect you from others and to make you aware of how your actions might affect your coworkers.

Most plumbers find themselves working around other people. This is especially true on construction jobs. When working around other people, you must be aware of their actions, as well as your own. If you are walking out of a house to get something from the truck and a roll of roofing paper gets away from a roofer, you could get an instant headache.

If you don't pay attention to what is going on around you, it is possible to wind up in all sorts of trouble. Cranes lose their loads sometimes, and such a load landing on you is likely to be fatal.

Equipment operators don't always see the plumber kneeling down for a piece of pipe. It's not hard to have a close encounter with heavy equipment. While we are on the subject of equipment, let me bore you with another war story.

One day I was in a sewer ditch, connecting the sewer from a new house to the sewer main. The section of ditch that I was working in was only about 4 feet deep. There was a large pile of dirt near the edge of the trench; it had been created when the ditch was dug. The dirt wasn't laid back the way it should have been; it was piled up.

As I worked in the ditch, a backhoe came by. The operator had no idea I was in the ditch. When he swung the backhoe around to make a turn, the small scorpion-type bucket on the back of the equipment hit the dirt pile.

I stood up when I heard the hoe approaching, and it was a good thing I did. When the equipment hit the pile of dirt, part of the mound caved in on me. I tried to run, but it caught both of my legs and the weight drove me to the ground. I was buried from just below my waist. My head was okay, and my arms were free. I was still holding my shovel.

I yelled, but nobody heard me. I must admit, I was a little panicked. I tried to get up and couldn't. After a while, I was able to move enough dirt with the shovel to crawl out from under the dirt. I was lucky. If I had been on my knees making the connection or grading the pipe, I might have been smothered. As it was, I came out of the ditch no worse for wear. But, boy, was I mad at the careless backhoe operator. I won't go into the details of the little confrontation I had with him.

That accident is a prime example of how other workers can hurt you and never know they did it. You have to watch out for yourself at all times. As you gain field experience, you will develop a second nature for impending coworker problems. You will learn to sense when something is wrong or is about to go wrong. But you have to stay alive and healthy long enough to get that experience.

Always be aware of what is going on over your head. Avoid working under other people and hazardous overhead conditions. Let people know where you are so you won't get stranded on a roof or in an attic when your ladder is moved or falls over.

You must also remember that your actions could harm coworkers. If you are on a roof to flash a pipe and your hammer gets away from you, somebody could get hurt.

Open communication between workers is one of the best ways to avoid injuries. If everyone knows where everyone else is working, injuries are less likely.

Primarily, think and think some more. There is no substitute for common sense. Try to avoid working alone, and remain alert at all times.

Chapter

5

The Basics of Plumbing and Related Science Issues

When you understand the basics of plumbing and related science issues, your perspective on being a plumber may change. The change should be for the better. Plumbers aren't scientists, but they do work with science on a daily basis. This doesn't mean you have to like science to be a plumber. The work you will do involves science and biology, but if nobody calls it by those names; it will just be plumbing.

The type of related sciences you apply to plumbing will not seem like science. For example, when you install a sewer and set the grade on the pipe, you are doing so because of the principle of gravity. As a plumber, you know pipes must run downhill to empty their contents, unless a pump is involved, but would you consider this knowledge scholarly? No, you would think of it only as plumbing. But you are working with gravity and scientific principles.

As another example, consider the venting of plumbing fixtures. As a plumber, you will install vents for most, if not all, plumbing fixtures. To you, this is a code requirement and standard procedure. The vent is just another plumbing pipe, but it is installed because of related sciences. The vents carry harmful gases to the outside air. In addition, the vents protect the fixture traps from vacuums and siphonage. The vents are working to accomplish goals based on science.

Remember the temperature and pressure relief valves we discussed for water heaters in Chapter 3? You, as a plumber, will know the device is required, and you will know it blows off if the water heater develops excessive temperature or too much pressure. Again, you are dealing with a related science issue. The factors involving the temperature and pressure fall into the science category.

As you can see from these examples, your work with science isn't going to be very difficult. It will come naturally through the understanding of plumbing. As you learn about hydraulics, bacteria, gases, gravity, siphonage, air properties, and so on, you will be learning about plumbing. When the lessons are put into the form of plumbing principles, they are easy to follow. There is no reason to be intimidated by the close connection of science and plumbing. Now, let's look at how your plumbing career will depend on related sciences.

The basics of plumbing are simple. Pipes bring potable water to a source for drinking and other pipes remove solid and liquid wastes. In its simplest form, almost anyone can accomplish plumbing. However, to serve all of its functions, plumbing gets a bit more complicated.

If you trace plumbing back to its origin, you will see that plumbers were working in ancient Rome. These plumbers worked with lead to form pipes that would serve their needs. Their pipes brought water to fountains, baths, and food crops. Drinking water was also conveyed in these early lead plumbing systems. Eventually, the pipes were used to move waste products to a point of disposal.

Since the days of the early Romans, plumbing has developed into a major industry. Hand pumps and outhouses are out, and high-tech fixtures are in. Plumbing has developed from the use of lead pipes and pitcher pumps to faucets that turn on automatically when hands are placed under the spout. The growth of plumbing has provided many financial opportunities, but it has also improved the standard of living for most people.

Can you imagine what New York City would be like without modern plumbing facilities? For that matter, can you imagine having to walk out to the spring to dip out your drinking water? Plumbing has come a long way, and much of its growth has happened in the last few decades.

It was not that long ago when people didn't have inside plumbing. When plumbing was moved inside, it remained somewhat primitive. Water still had to be pumped in by hand, and toilet facilities were often drained to a pit outside the house. These improvements made life easier, but they still left a lot to be desired.

As time passed, septic systems were brought into use and electric pumps replaced the hand pumps. Plumbing was on a roll, and it is still rolling along today. The longer we study plumbing, the more we learn. The more we learn, the more we change. For example, it wasn't many years ago that plumbers used solder containing 50 percent lead to make potable water pipe joints. This isn't done today, at least not in compliance with the plumbing codes. The solder with heavy lead content is no longer allowed for use with potable water systems.

Why was 50/50 solder banned? It was decided that the lead in the joints may have an adverse effect on public health. Plumbing is constantly changing to meet higher health standards. Some of these changes include backflow preventers, vacuum breakers, and more efficient venting systems. They all reflect related sciences.

Plumbing provides more than convenience; it provides sanitary living conditions. Piping is used for many purposes. Plumbers use piping to carry both potable and waste water. The gas piping normally done by plumbers contains natural or propane gas. Heating technicians use piping to convey nonpotable water and steam through heating systems. They also use piping to bring fuel to the heating systems. Refrigeration mechanics use piping to transport chilled water. With so many trades using piping for so many purposes, one has to be careful not to make mistakes with cross-connections.

Cross-Connections

Cross-connections present dangerous health risks. Can you imagine what would happen if a plumber tapped into a heating pipe, thinking it was a potable water pipe? It could happen, very easily. Most residential heating systems that covey hot water use ¾-in copper tubing. This type of piping is the same as that used for much of the potable water distribution in homes. The tubing used for both purposes is copper, so it all looks the same. An inexperienced plumber could tap into a heating line by mistake and bring dirty boiler water into the potable water system.

If this were to happen, all kinds of problems are possible. Boiler water gets very hot. A person who stepped into the new shower and cut on the hot water might be scalded or might ingest harmful bacteria from the kitchen faucet.

Once the cross-connection is made, the nonpotable water from the heating system is allowed to mix with the potable water. Even though the mistake would probably be noticed quickly when the dirty boiler water came out of a faucet, damage would already be done. The potable water pipes would be contaminated. Cross-connections can ruin good plumbing systems and endanger public health and safety.

There are other ways for cross-connections to contaminate potable water sources. It doesn't take a blatant mistake, like tapping into the wrong pipe, to create a hazard. Backpressure and back siphonage are two other ways to contaminate a system through cross-connections.

Backpressure

Backpressure can push contaminated substances into a pipe or fixture meant for sanitary use. In other words, backpressure can push sewage or gray water up the plumbing pipes until it overflows into a sanitary fixture, like a sink, lavatory, or bathing unit. If your work will involve drain cleaning, you will see plenty of this type of problem.

If a septic tank has reached its capacity or there is a blockage in a pipe, waste in the pipe begins to back up. As the volume of waste in the pipe increases, it has nowhere to go. Therefore, it comes up in the toilet, tub, or sink. This example may not be the best one to describe backpressure, but it does give you a good mental picture of what happens when backpressure occurs.

Atmospheric pressure is at the root of backpressure. When the pressure of the nonpotable pipe is higher than that of the potable outlet, the problem of backpressure occurs. Have you ever flushed a toilet and heard a gurgling in the lavatory or bathtub? If you have, you heard backpressure. Proper venting greatly reduces the chance of a backpressure problem.

Back Siphonage

Back siphonage is another way cross-connections contaminate potable water systems. Back siphonage occurs when a vacuum is formed that pulls nonpotable water into the potable water system. These accidents can happen in many different ways. If a pipe breaks, it can create siphonage. The broken pipe can cause contaminated substances to be sucked into the potable water system through hoses and other unprotected devices. Back siphonage could result from a common activity, such as replacing a water heater. Let's look at one way a home's potable water system might be contaminated with deadly chemicals.

In this example the plumbing in the house is not equipped with backflow preventers and vacuum breakers. The homeowner is out in the flower garden. She is spraying her garden with pesticides from a sprayer that connects to a garden hose. This is the type of sprayer that has a large reservoir for chemicals and depends on the water pressure from the garden hose to disperse the chemicals. The woman has just refilled her sprayer with chemicals when her son calls her to the house. She lays down the sprayer and goes in the house.

The homeowner's son explains how he tried to take a shower and there was no hot water. The lady tries the sink faucet, and there is no hot water to be had. She calls a plumber.

The plumber arrives and finds the water heater should be replaced. The homeowner authorizes the work, and the plumber begins by draining the old water heater. The plumber's apprentice goes back to the shop and picks up the replacement water heater and the required plumbing permit. Meanwhile, the plumber has drained the water heater and cut the inlet and outlet pipes. The water had to be cut off at the main valve because the valve at the water heater needed to be replaced. It was a stop and waste valve, and code requires a positive-closing valve, like a gate valve.

The new water heater is installed and inspected. Nobody is aware that the potable water system has been poisoned. What happened? Cutting the pipes at the water heater created siphonage that pulled water out of the pipes above the water heater. In draining the plumbing system, the plumber unknowingly poisoned it by siphoning the chemicals out of the sprayer, through the garden hose, and into the pipes.

This type of accident could easily happen. The plumber didn't do anything wrong and he had no idea there was a chemical sprayer attached to an open silcock. The woman never thought about the chemicals once her mind was diverted to the replacement of her water heater. The results of such an accident could be very bad. They could also be avoided very simply. If the silcock had been equipped with a vacuum breaker, the accident wouldn't have happened.

As I hope you are starting to see, the science in plumbing isn't hard to understand if you look at it properly. And code requirements for backflow prevention exist to protect the public, not to make life hard for plumbers or jobs expensive for consumers.

Water

We all need water to survive. Plumbers help to bring water to the public. They do so by installing piping and pumps. But getting water to the faucet isn't always enough. The water used for drinking must be potable—safe to drink.

Since water is accumulated in various ways, there are many options available for obtaining it. Water can come from underground veins. This underground water is accessed by digging, drilling, boring, or driving a well shaft into the earth. Once the shaft connects with the underground water source, a pump is normally used to lift the water above ground. The pump also gets the water to the desired fixture locations. This is once again a matter of science. The pumps use electricity, centrifugal force, and pistons to pump the water.

Other sources of water include rivers, lakes, streams, oceans, springs, and cisterns. All of this water can be pumped to a plumbing fixture but much of it is not potable. Therefore, the water must be treated before it can be consumed safely by humans.

People who get their water from municipal water supplies are drinking conditioned water. Municipalities treat their water supplies in various ways. Filtering removes mud and solid matter. Aeration removes gases that may be present in the water, gases like carbon dioxide, hydrogen sulfide, and carbon monoxide. Chemicals, like chlorine, are used to kill bacteria. There are numerous ways in which water can be treated for safe drinking. If you will notice, we are talking about the removal of gases and bacteria, another form of science.

Are you thinking that plumbers don't have to be concerned with water treatment since the municipalities take care of it? If so, you are wrong. Many plumbers work in rural areas where homes and businesses derive their drinking water from private wells. It is common for plumbers to install water conditioning equipment in these private residences and businesses for water conditioning.

Plumbers who deal in water conditioners must know something about how they work. For example, well water often contains high contents of calcium or magnesium. When this is the case, the water is considered to be hard water. Hard water can affect the operation of plumbing fixtures, and it is difficult to clean clothes, dishes, or human hair in hard water. Because of this, water softeners are often installed by plumbers to counteract the problem.

Private water sources can also suffer from high contents of acid. Acidic water can eat holes in copper piping, and it can affect people's health. Acid neutralizers, installed by plumbers, can eliminate the bad effects of acidic water. Are you starting to see how plumbers work with science regularly, sometimes without even realizing it?

Waste

The removal of waste is one of the most important jobs of plumbing. The sanitary removal and disposal of sewage is essential to good health. Today's plumbing systems use septic systems for private waste management and municipal facilities for public waste control. Municipal facilities treat waste water to a point where it is safe to discharge back into the environment. Septic systems do essentially the same job.

Before waste and sewage can be returned to the earth safely, the chemicals in it must be rendered harmless. This is done in different ways and usually requires many steps. For example, lets look at how a municipality might treat its waste water.

The water treatment plant might use a filtering process to clean up its waste water. The waste and sewage will go through a screen and grit chamber to remove debris, like sand and trash. When the waste gets to the holding tank, it will remain there until the solid waste settles to the bottom. The solid waste is called sludge and the dirty water is called effluent.

The effluent will be run from the holding tank to a location where it will be aerated. The aeration allows the chemicals to be broken down into a harmless stage. When the chemicals are no longer a threat, the water can be put back into the environment.

Some cities use different methods for treating their waste water. They still filter the waste and sewage, but they induce bacterial growth to break down the waste. Bacteria plays a big role in treating waste and sewage. Aerobic bacteria dissolves materials with the use of oxygen. Anaerobic bacteria doesn't rely on oxygen, but it can produce unpleasant odors.

Septic systems, like those used for residences, also rely heavily on bacterial action. Waste and sewage enters the septic tank from the home's sewer. Once in the tank, the solids sink to the bottom and effluent remains near the top of the tank. As the capacity of the tank is reached, the effluent is passed out through a pipe and into a distribution box. From the distribution box, the effluent is sent out to the septic field. The sludge that remains in the septic tank must be pumped out periodically and trucked to a proper disposal site.

Once the effluent enters the septic field, it runs through slotted pipes that allow it to disperse into the ground. As the effluent passes through the earth, it is aerated and cleaned. The ground containing the septic field must meet certain percolation requirements; otherwise, the effluent would not be properly treated before reentering the ground water system.

Isn't it amazing how much plumbers work with the various sciences in their daily jobs? Would you have ever considered that there were armies of good bacteria at work in a septic tank? Plumbers work with science all the time. Many apprentices become apprehensive when they find they will have to learn related sciences as a part of their training. Visions of strange formulas and weird professors come to their minds. As you can see, you will work with science, but there is not much need for fearing it.

Gases

Gases are all around us. We breath them, we set them on fire, and we use them to power tools and appliances. Gases can be found naturally, and they can be made through chemical processes. There are several useful gases employed in the plumbing trade. There are also some gases that pose particular threats to plumbers.

Acetylene is often used by plumbers to fuel torches. Propane is also used for torches, and it can be used to fuel heating systems and appliances, as well. Butane is sometimes used as a heating fuel. Oxygen is often mixed with acetylene to produce intense heat for welding. Chlorine gas can be used in the purification of water. Sewer gas is hydrogen sulfide. Methane gas is also formed in sewers and septic tanks, and it is explosive when confined. Ammonia can be found in refrigeration and cooling systems. Natural fuel gas, made mostly of methane, is used to fuel heating systems and appliances. Plumbers can come into contact with all of these gases, and more.

Some gases are particularly dangerous. A few of the gases to be especially careful of include chlorine, sulfur dioxide, carbon monoxide, hydrogen, methane, and ammonia.

Plumbers need to be aware of all the various gases that can present problems. Many gases are formed as the result of some work activity. For example, the use of plastic solvents and flames can create harmful gases. Raw sewage produces gas that can explode. Leaving a truck running to stay warm will produce carbon monoxide. There are plenty of ways to do yourself in with gases if you are not careful.

Well, there you have it; we are done with the subject of related sciences. Of course, there is always more to learn, but now you have the basics. We are now ready to move on to the next chapter and see what types of tools you will need.

Chapter

6

Tool Selection

Tool selection can become confusing for apprentices. There are hundreds of possibilities for what should go in a plumber's toolbox. Some tools are a given but many have only specialized uses and may not be required. However, some specialty tools, like a basin wrench, are mandatory equipment.

Most plumbing companies provide their employees with the big tools that will be used in the trade. The tools supplied by employers might include right-angle drills, reciprocating saws, sewer machines, jackhammers, and the like.

Individual plumbers and apprentices have to supply their own hand tools. These tools cover a lot of options. Some of the tools you can't get by without are tape measures, hammers, screwdrivers, pliers, and wrenches. Torches are generally provided by employers but not always.

Tools are a big issue in plumbing. They are used daily, and they can make a plumber's life much easier. There are many plumbers who prefer to buy their own tools, even if the company is willing to provide them. These plumbers want a certain brand or type of tool that may be different from what the company offers. Some plumbers are very particular about their tools.

Then you will find the plumbers who try to get by without buying any decent tools. They will use whatever the company gives them and try to borrow tools from their coworkers. When these plumbers do buy a tool, it is usually the cheapest version available. Plumbers who have inadequate tools are limited in their abilities to be good plumbers.

In this chapter, we are going to look at just about every tool a plumber ever uses. By the time you are done reading this section of the book, you will know what tools are needed and what they are used for. Along the way, you will be given advice on when to buy top-quality tools and when you can get by with lower-quality tools.

Typically, I recommend buying top-notch tools. They are an investment you will use for many years, and having the right tool will make your job more enjoyable. However, there are times when I lean toward the no-name tools.

When I entered the plumbing trade, I bought about $500 worth of tools. The price included a right-angle drill and a reciprocating saw. The toolbox I bought was a big-name brand and so were most of the tools. I might add, I had to borrow the money for the tools.

Since I was young and didn't know much about tools, I asked my father to help in the tool acquisition. His help proved valuable, and his advice was good. We did, however, butt heads on the issue of a socket set. He wanted me to buy a high-dollar set of quality sockets. I opted for a cheap set in a plastic box. He gave me the lecture on how I was wasting my money on the cheap sockets. Quoting experiences from his past in the automotive industry, he gave more reasons than I could count for my great mistake.

Well, at the time, I couldn't prove him wrong. I was just a wet-nosed rookie who didn't know beans about tools. Even so, I stood my ground and kept the little orange box of sockets. You see, I didn't think plumbers used sockets too often, so I felt comfortable in my decision.

Since that day when we drove 50 miles to the big-city hardware store, some 18 or 20 years have passed. Many of these years found me out in the field, working with my tools. Today, I still have several of my original tools. They are not hung on the wall as memorabilia; they are in the tool box, ready for use. My father came to visit recently, and we found a little odd job to do around the house. Surprisingly, the job required the use of a ratchet and socket. I went to the basement and returned with my battle-scarred orange box, still full of cheap sockets. As we worked a little and talked a lot, I kidded Dad about the sockets. Those cheap sockets have spent most of their days in my truck, and they are still going strong. Admittedly, they didn't see a lot of regular use because the plumbing I worked on didn't require sockets often. Well, we had a big laugh and a great visit.

The point to this piece of my personal history is that my el-cheapo sockets were a solid investment. They always got the job done, and they cost a fraction of what the good sockets would have set me back. Looking back, I wish I hadn't spent so much money on the toolbox. Anyway, as we go along, I will give you more advice from my experience about what tools you should buy to start with and what tools to work your way up to.

There is another good reason for buying inexpensive tools. Tools get stolen and lost. If you will be working around an unsavory crowd, it doesn't pay to leave expensive tools laying around. When you become a plumber, you have to monitor your apprentice closely. Helpers are famous for leaving tools under houses and scattered about the job. Until you pay for your own tools, it is harder to respect them. Once your hard-earned money is invested, your memory gets better about finding and picking up your tools.

As a word of advice, keep your plumber's tools in mind when you get in the field. It will probably be your responsibility to pick up at the end of the day, and if you leave your plumber's tools behind, your work the next day will almost always be harder.

Tool Boxes

You should start your tool selection with a toolbox (Fig. 6.1). You will need the box to keep your new tools in. But, when you start to look at tool boxes, you may be amazed at the various sizes and styles available. There are tall ones, short ones, long ones, and fat ones. Some boxes have removable trays and some have hinged trays that fan out. There are metal boxes and plastic boxes. Which one will you buy?

As an apprentice, you are not going to need a lot of tools. Until you get your plumber's license, you can get by with one toolbox of moderate size. As your tool collection grows, you will want several boxes. You will have one for tools that are rarely used, one for tools kept close at hand, and one for the tools you use on most every job. Many seasoned plumbers use a 5-gallon bucket for their every-job tools.

The bucket is handy for quick access, but it is an unorganized way to tote your tools. However, the bucket can be emptied and used to catch dripping water. It can be turned over and used as a seat, and some plumbers use them as miniature stepladders. I will admit that I sometimes use buckets to stand on, but you shouldn't. Buckets aren't step stools, and they shouldn't be used as such.

Soft-sided bags are another option for carrying your tools. My tools are in a bucket, a metal tool box, a plastic tool box, and a soft-sided bag. Each container carries different types of tools. This arrangement works well for me, but you can carry your tools in any type of container you like.

Figure 6.1 Tool box. (*Ridge Tool Co.*)

For a starter box, I would choose a strong plastic box with a removable tray. Plumbers' boxes get wet, and metal boxes rust. Durable plastic boxes stand up to a lot of abuse. As for size, most any medium size box will do. A box that is about 2 feet long and about 10 inches high will do nicely.

Wrenches

Pipe wrenches

Pipe wrenches (Fig. 6.2) come in various sizes and are found on every plumbing truck. These wrenches are not used as much as they used to be, but they still see frequent use in modern plumbing applications.

Pipe wrenches are used for installing threaded pipe, like gas pipe. They are used to hold and turn large pipes, fittings, and nuts. When doing remodeling work, pipe wrenches are used for repairing and removing old piping. They come in handy on service calls, and they are used for installations of gas piping, sprinkler systems, heating systems, and so forth.

Pipe wrenches range in size from very tiny to extremely large. You have a choice of steel or aluminum for the frame construction of your wrenches. The aluminum wrenches are more expensive than the steel ones, but they are much lighter. The lightweight wrenches are my favorite.

What sizes do your wrenches need to be? As an apprentice, you may not be required to have pipe wrenches. If, however, you need wrenches, you should consider the following sizes: 8, 10, 12, 14, and 18 inches. If you are strapped for money, go for the 10-, 14-, and 18-inch wrenches.

Don't skimp on quality with pipe wrenches. Cheap wrenches have sloppy jaws and substandard teeth.

Offset pipe wrenches

Offset pipe wrenches (Fig. 6.3) are specialty tools. They work well when the job involves using the wrench over your head or in tight spots. As an apprentice, you shouldn't need offset wrenches. When you are out on your own, you might find these wrenches helpful. If you are going to buy an offset pipe wrench, buy a good one.

Chain wrenches

Chain wrenches (Fig. 6.4) are also specialty tools. They are used in place of pipe wrenches when working space is extremely limited. The chain is wrapped around the pipe, fitting, or nut and locked into place on the handle. At that point, the wrench can be used to turn the object. Chain wrenches are put under a lot of stress. When you decide to add a chain wrench to your tool box, buy a good one.

Figure 6.2 Pipe wrench. (*Ridge Tool Co.*)

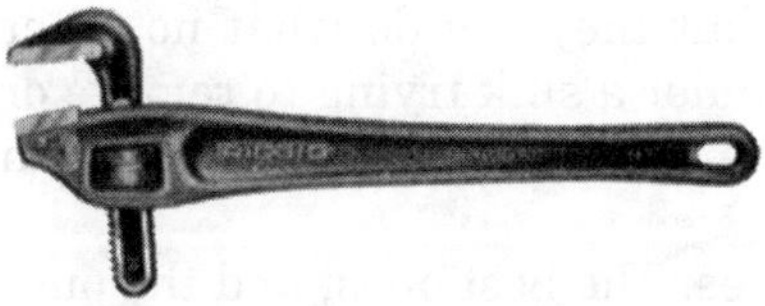

Figure 6.3 Offset pipe wrench. (*Ridge Tool Co.*)

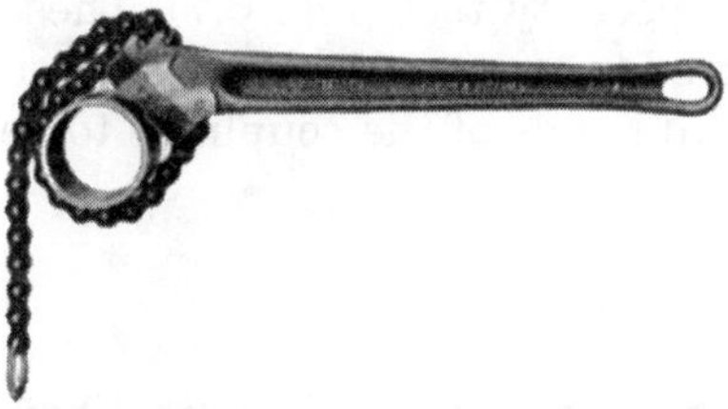

Figure 6.4 Chain wrench. (*Ridge Tool Co.*)

Strap wrenches

Strap wrenches (Fig. 6.5) work on the same principal as chain wrenches. However, they incorporate the use of a web strap rather than a chain. The strap doesn't give you as good a grip, but it doesn't scar the material you are working with either.

A tub spout is a good example of where a strap wrench would be a better choice than a chain wrench. Since the tub spout is in plain sight

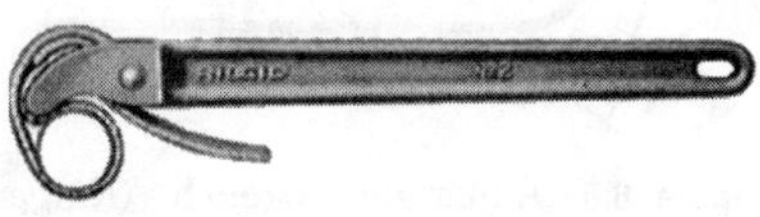

Figure 6.5 Strap wrench. (*Ridge Tool Co.*)

and made of a finished material, it would not be wise to use a chain wrench to loosen or tighten it. The strap wrench would not cut the spout, but the chain would.

Strap wrenches are also under stress when used. If you are going to buy a strap wrench, go for quality.

Basin wrenches

Basin wrenches (Fig. 6.6) are essential to plumbers and apprentices. These wrenches look a little strange, but they can do what no other wrench can. When you are curled up under a sink trying to remove or install a faucet, you will quickly see the importance of having a basin wrench.

Basin wrenches come in different sizes. The best ones, and the ones I would recommend, are able to telescope out to varying lengths. These specialty tools see a lot of use and abuse; buy a good one.

Torque wrenches

Torque wrenches are another type of specialty tool (Fig. 6.7). These tools are used with the couplings installed to connect hubless cast-iron pipe. They allow the clamps of the metal bands of the couplings to be tightened to a measured torque.

Adjustable wrenches

Adjustable wrenches (Fig. 6.8) should be in every plumber's tool box. They are used frequently by both plumbers and apprentices.

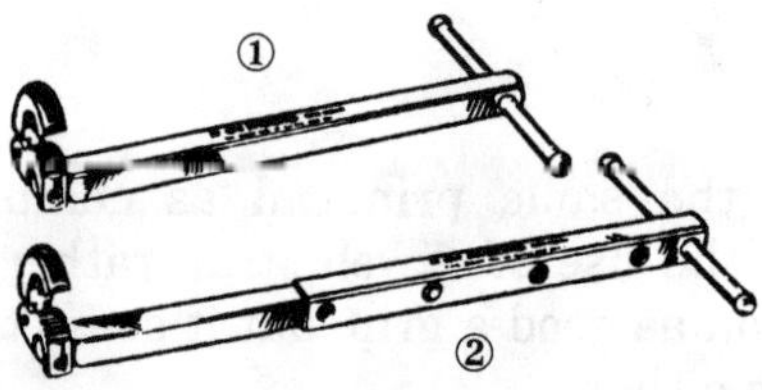

Figure 6.6 Basin wrench. (*Ridge Tool Co.*)

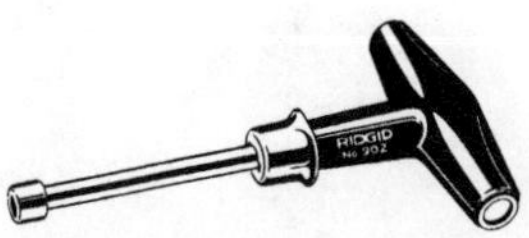

Figure 6.7 Torque wrench. (*Ridge Tool Co.*)

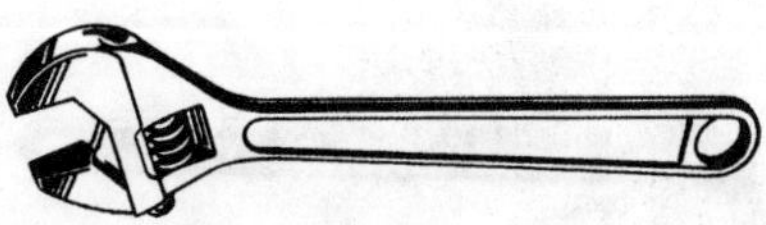

Figure 6.8 Adjustable wrench. (*Ridge Tool Co.*)

Figure 6.9 Tub wrenches.

Tub wrenches

Tub wrenches (Fig. 6.9) are needed to remove faucet stems from many types of tub and shower valves. Without these wrenches, removal of the stems is all but impossible.

Saws

Jig saws

Jig saws (Fig. 6.10) are not used by plumbers often. When they are used, they are normally used to cut holes in countertops for sinks and

Figure 6.10 Jig saw. (*Ridge Tool Co.*)

lavatories. You won't need a jig saw as an apprentice, and I doubt if you will use one even after you are a licensed plumber. Normally, carpenters cut the holes needed in counters. Since jig saws are not used much, you can buy an inexpensive model.

Circular saws

Circular saws (Fig. 6.11) do see some use by plumbers but not a lot. A few plumbers use circular saws to cut pipe. In general, circular saws are not all that popular with plumbers. Because of the lack of use, an inexpensive circular saw will get the job done for a plumber.

Hacksaws

Hacksaws (Fig. 6.12) are another primary tool for plumbers and apprentices. There are countless times when you will rely on your hacksaw. It is used to cut pipe, bolts, nuts, and a host of other objects.

There is a type of hacksaw that allows the blade to be mounted in several positions. The blade can also be extended straight out from the front of the saw. This allows you to make cuts even when the frame of the saw won't fit in the cutting space. Another option is to mount the blade on an angle. This makes cutting closet bolts and similar items easy and efficient. It also reduces the risk of scarring fixtures with the frame of the saw.

Buy a top-quality hacksaw, and don't try to save money on blades. The hacksaw is only as good as its blades.

Universal saws

Universal saws (Fig. 6.13) are favored by some plumbers for cutting plastic pipe. The saws have replaceable blades and are meant for cutting pipe. They will, however, cut wood and other objects. Universal saws are not expensive, and they are all in the same price range. I prefer to use my hacksaw for cutting pipe.

Jab saws

Jab saws (Fig. 6.14) are specialty items. They consist of a round handle and a blade, normally a long one. These saws allow you to cut in areas that are inaccessible to other saws. This is not a tool you have to have, but it will be welcome under special conditions. Because of the nature of the saw, it is best to buy a quality unit.

Reciprocating saws

Reciprocating saws (Fig. 6.15) see almost as much use in the plumbing trade as right-angle drills. These versatile saws are used for cutting

Figure 6.11 Circular saw. (*Ridge Tool Co.*)

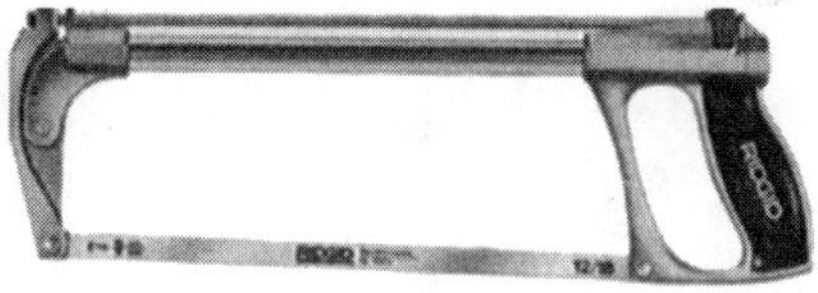

Figure 6.12 Hacksaw. (*Ridge Tool Co.*)

Figure 6.13 Universal saw. (*Ridge Tool Co.*)

Figure 6.14 Jab saw. (*Ridge Tool Co.*)

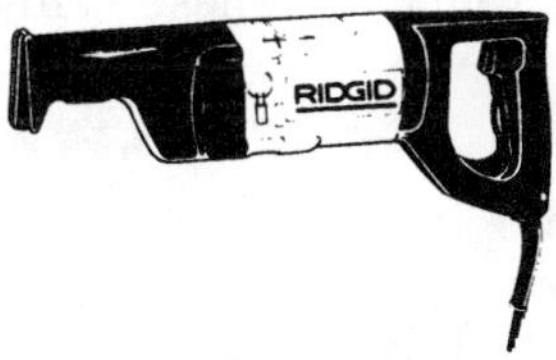

Figure 6.15 Reciprocating saw. (*Ridge Tool Co.*)

everything from pipe to plywood. With an assortment of quality blades, there isn't much you can't cut with these saws. Like right-angle drills, most employers provide their employees with reciprocating saws. Since these saws see frequent use, buy the best.

Drills

Right-angle drills

Right-angle drills (Fig. 6.16) are have-to-have tools for plumbers. These drills are sometimes used every day. There is no reasonable way that a residential plumber can get through most installations without a right-angle drill.

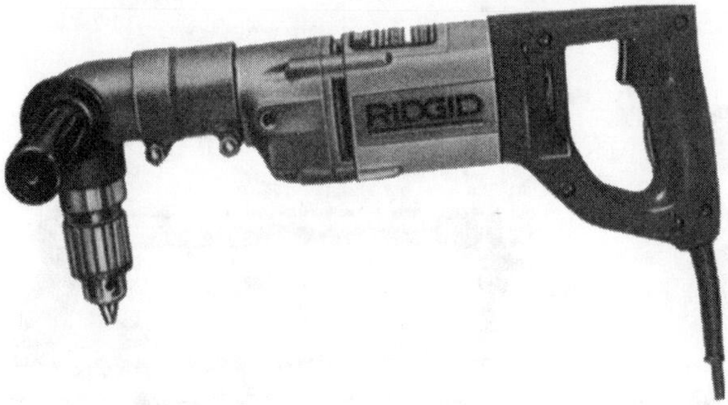

Figure 6.16 Right-angle drill. (*Ridge Tool Co.*)

As an apprentice, you will probably get to know these drills very well. However, you shouldn't have to buy your own. Even as a plumber, your employer should furnish you with a drill and bits. If you decide to buy your own drill, buy the best you can afford.

Hand drills

Small hand drills (Fig. 6.17) are sometimes used by plumbers. If you want a hand drill, buy one with a variable speed control. The drill should be chucked for ⅜-inch bits. Since you won't use these drills often, an inexpensive one will get the job done.

Torches

Air-propane torches

Torches are used by all plumbers, and a lot of plumbers prefer air-propane torches (Fig. 6.18). Propane torches are fine for any normal plumbing application. Even the little hand-held propane torches home-

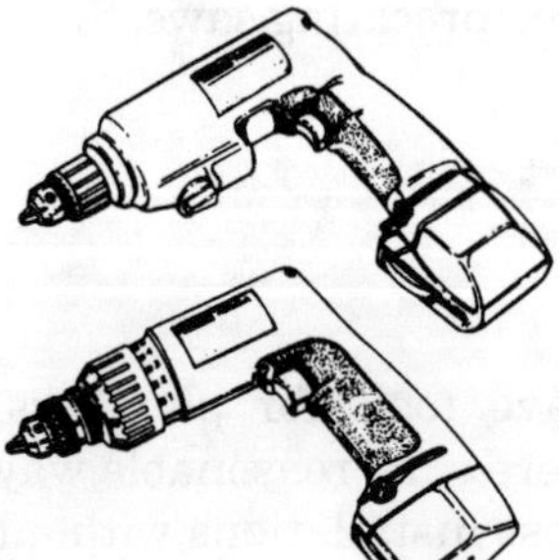

Figure 6.17 Hand drill. (*Ridge Tool Co.*)

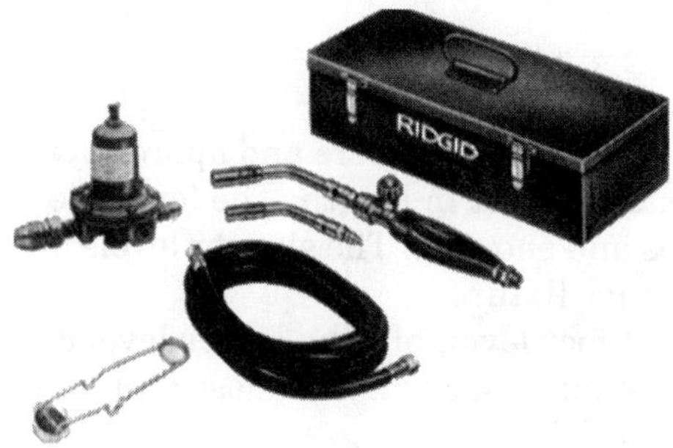

Figure 6.18 Air-propane torch. (*Ridge Tool Co.*)

owners favor can get many plumbing jobs done. These small torches are particularly useful when you have to crawl around under houses. Torches are normally supplied by employers. Advanced apprentices do a lot of soldering, so you might want your own torch, but you shouldn't be required to purchase one. If you purchase a torch, get a good one.

Air-acetylene torches

Air-acetylene torches (Fig. 6.19) are the other type of torch used by plumbers. These torches produce hotter flames and have been around for a long time. Until you use both types of torches, you won't be able to decide which you prefer. Personally, I prefer acetylene torches. Remember to buy a good torch, if you buy one.

Torch strikers

Torch strikers are simple, inexpensive tools used to light torches. Every apprentice should own a striker.

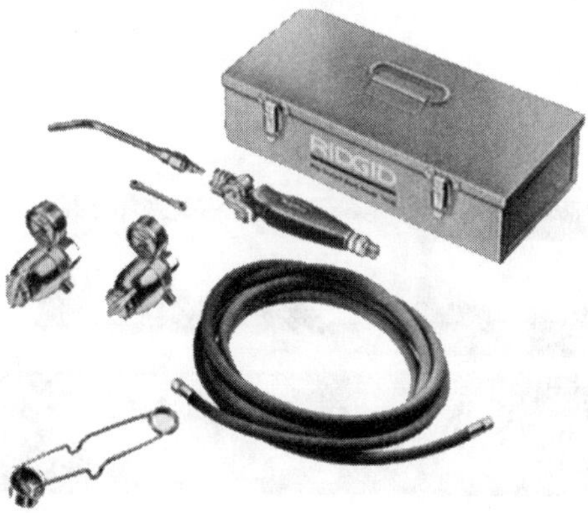

Figure 6.19 Air-acetylene torch. (*Ridge Tool Co.*)

Levels

Standard levels

Standard levels (Fig. 6.20) are used by both plumbers and apprentices. Most of the levels used are either 2 or 4 feet in length. The longer levels are used when setting bathtubs and showers. The shorter levels are used when installing toilets and other fixtures.

It would be a good idea to get a 2-foot level, but the larger level can wait until you are licensed to work on your own. For most plumbing jobs, an inexpensive level will do nicely.

Grade levels

Grade levels (Fig. 6.21) are specialty tools that most plumbers can get by without, but they help make a lot of jobs go faster and better. Most grade levels are 2 feet long and have a thumb-wheel at one end. The thumb-wheel is used to raise and lower a threaded stud. The stud can be adjusted to make grading pipes much easier.

Most plumbing pipes are installed to drop ¼ inch for every foot they travel. In other words, when a 4-foot piece of pipe is installed, the downhill end of the pipe should be 1 inch lower than the uphill end. Grade levels make this type of pipe grading simple. If you have a 2-foot grade level, you lower the stud so that it extends a half an inch below the bottom edge of the level. When you place the level on a sloping pipe, with the stud on the downhill end, the bubble in the level should read level. If it does, the pipe is properly graded for a ¼-inch-to-the-foot fall.

Most apprentices don't own grade levels, but if your work will involve installing a lot of drain and vent pipes, you should buy one as soon as you can afford it. Most grade levels are all about the same price.

Figure 6.20 Standard level. (*Ridge Tool Co.*)

Figure 6.21 Grade level. (*Ridge Tool Co.*)

Torpedo levels

Torpedo levels (Fig. 6.22) belong in every plumbing toolbox. These levels are used all the time. They are used to set the degree of turn for fittings and to spot check pipe grade and fixtures, among other things. Torpedo levels don't cost much, but they are great. You have to have a torpedo level. Some models have a magnetic strip on one edge. This allows the level to be stuck to metal and cast iron pipe. Since most drainage pipe used today is nonmetallic, the magnetic strip isn't as useful as it once was. Good torpedo levels have the ability to set fittings at angles. You will use this type of level often and should buy a good one.

Line levels

Line levels (Fig. 6.23) are used by plumbers installing underground plumbing. A string is pulled tight above the work area, and the level hangs on the string. This allows the string to be positioned in a level manner. The string can represent the height of a concrete floor, or it can be used to check the grade of a pipe. Line levels are inexpensive, but most apprentices won't need them. There is not much point in buying an upper-end line level.

Screwdrivers

Standard screwdrivers

Standard screwdrivers (Fig. 6.24), in various sizes, are used frequently by plumbers and apprentices alike. You will need an assortment of standard screwdrivers. Some should have long handles. Others should

Figure 6.22 Torpedo level. (*Ridge Tool Co.*)

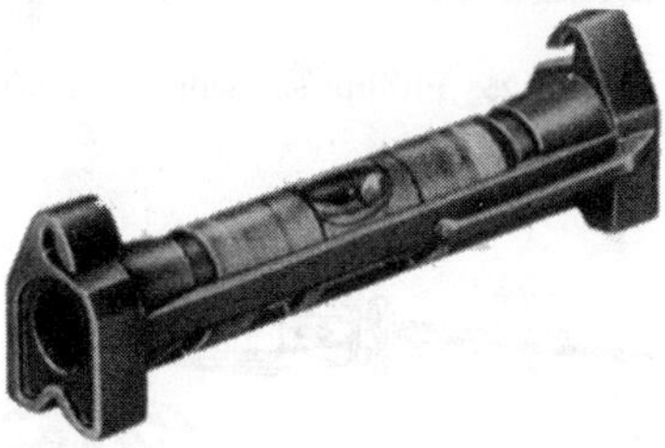

Figure 6.23 Line level. (*Ridge Tool Co.*)

Figure 6.24 Standard screwdriver. (*Ridge Tool Co.*)

be fitted with short, medium, and stubby handles. There are many conditions under which the screwdrivers will be used, and varying lengths and bit sizes will be needed. All of the screwdrivers you purchase should be of good quality. Cheap ones can ruin screws or twist or break.

Phillips-style screwdrivers

Phillips-style screwdrivers (Fig. 6.25) fall into the same category as standard screwdrivers. Buy an assortment, and remember, only use screwdrivers as screwdrivers; they are not chisels.

Combination screwdrivers

Combination screwdrivers (Fig. 6.26) are very popular in the plumbing trade. These screwdrivers have shafts that pull out of the handle and accept different sizes and types of bits. Normally, there will be two standard bits and two Phillips-type bits.

One of these screwdrivers can accomplish most of your plumbing tasks. However, they can't replace all of your other screwdrivers. Buy a combination screwdriver, but also get the other screwdrivers.

Awls

Awls (Fig. 6.27) see occasional use in plumbing but not much. Awls are used for punching and lining up holes. They are not a bad tool to have, but you shouldn't need one. If you buy an awl, buy one with good metal in it.

Figure 6.25 Phillips screwdriver. (*Ridge Tool Co.*)

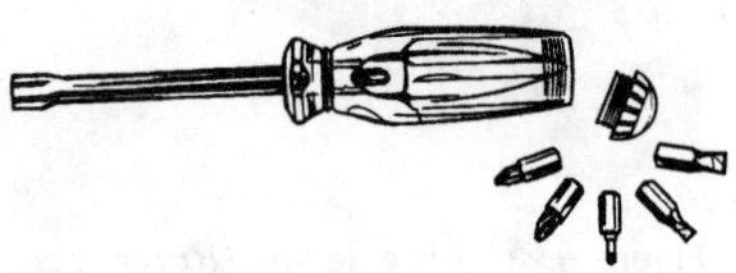

Figure 6.26 Combination screwdriver. (*Ridge Tool Co.*)

Figure 6.27 Awl. (*Ridge Tool Co.*)

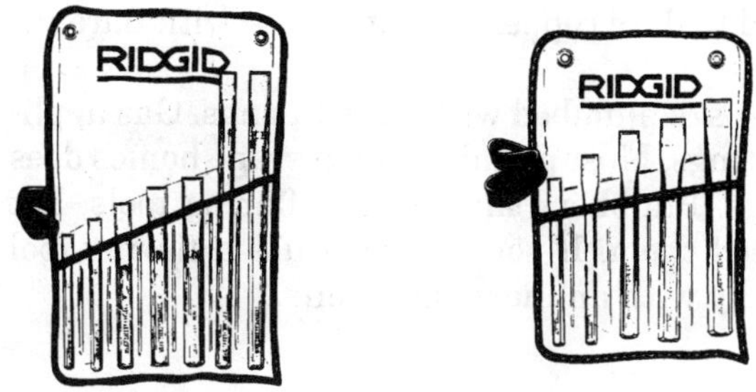

Figure 6.28 Cold chisels. (*Ridge Tool Co.*)

Chisels

Cold chisels

Cold chisels (Fig. 6.28) are used to chip concrete and to make holes in masonry walls. Apprentices usually do this work. Therefore, equip yourself with a selection of cold chisels in various sizes. Spend enough money on these tools to get the quality you need for durable service. Cold chisels take a pounding, literally, and need to be strong.

Wood chisels

Plumbers use wood chisels a lot more than most people would think. The chisels are used to shave band boards, expose nails, and similar jobs. You should buy an assortment of wood chisels. Ideally, you should have at least three chisels in different sizes: small, medium, and large. The chisels don't have to be of superior quality, but you must keep them sharp.

Flaring Tools

Flaring tools (Fig. 6.29) are used frequently by some plumbers and seldom by others. It all depends on the type of work that is being

Figure 6.29 Flaring tools. (*Ridge Tool Co.*)

done. Flaring tools are used to flare copper tubing for use with flare fittings.

Many older mobile homes are plumbed with flare fittings. Gas appliances often rely on flare fittings. Most plumbing in average homes does not depend on flare fittings. Plumbers should have flaring tools, but most apprentices won't need them. If you will only use a flaring tool now and then, you can get by with an inexpensive one.

Spring-Type Tubing Benders

Spring-type tubing benders (Fig. 6.30) can make life a little easier for apprentices. These devices are used to bend sensitive supply tubing without kinking it. Experienced hands won't need spring benders, but when you are first starting out, they will keep you from ruining supply tubes. Prices on spring benders are pretty uniform.

Ladders

Step ladders

Step ladders (Fig. 6.31) get a lot of use in the plumbing business, but they are generally provided by the employer. However, if you are buying your own ladder, buy the best you can afford. Cheap ladders are dangerous and don't last long. This advice also applies to extension ladders.

Extension ladders

Extension ladders (Fig. 6.32) are not used nearly as often as step ladders in plumbing, but they do have their place. Your employer should supply any ladders you need.

Shovels

Ah, shovels. Apprentices get to know shovels (Fig. 6.33) very well. You will probably develop a deep relationship with your shovel, but your employer will supply your digging tools. Shovel quality is not too important, as long as the handle is strong.

Figure 6.30 Tubing bender. (*Ridge Tool Co.*)

Figure 6.31 Step ladder. (*Ridge Tool Co.*)

Figure 6.32 Extension ladder. (*Ridge Tool Co.*)

Figure 6.33 Shovels. (*Ridge Tool Co.*)

Hammers

Hammers will be needed during your apprenticeship. You will need a claw-type hammer (Fig. 6.34), like carpenters use. The claws can be curved or straight. The face of the hammer can be smooth or patterned. I prefer a heavy framing hammer, with straight claws and a patterned face.

A small sledge hammer is also a good investment. It will make your work with cold chisels easier and faster. A hammer in the 2-pound class is fine. Hammers are used frequently, and you should look for quality in these tools.

Pipe Extractors

Pipe extractors (Fig. 6.35) are used to remove threaded sections that have broken off in a fitting. Apprentices don't need these tools, but plumbers do need them from time to time. There is no need wasting money on cheap extractors. If you buy them, buy good ones.

Figure 6.34 Claw hammer. (*Ridge Tool Co.*)

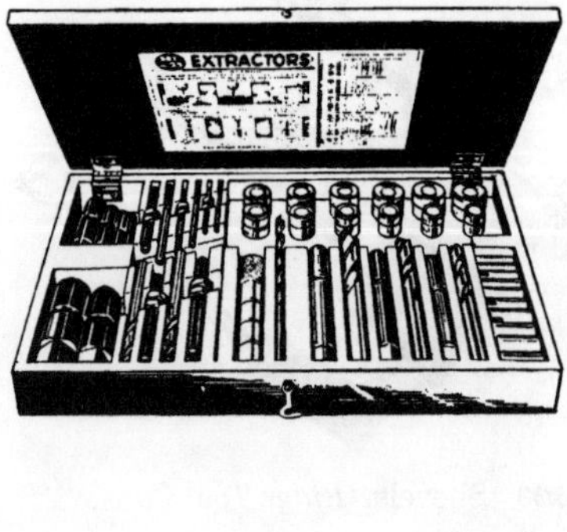

Figure 6.35 Extractor set. (*Ridge Tool Co.*)

Cutters

Bolt cutters

Bolt cutters (Fig. 6.36) are used in plumbing mostly to cut rebar in concrete. Unless you are involved in a lot of remodeling work, you shouldn't need bolt cutters. On the few occasions you need them, your employer should provide them. Bolt cutters do a tough job. If you are going to buy them, buy good ones.

Standard tubing cutters

Anyone working in the plumbing trade needs standard tubing cutters (Fig. 6.37). These devices are used to cut copper tubing. While it is possible to cut copper tubing with a hacksaw, the cut edges will be rough and uneven. Standard tubing cutters, also called roller cutters, cut pipe and tubing with a smooth, uniform edge. They are also much easier to use than a hacksaw.

These cutters come in many sizes. For residential plumbing, the largest copper tubing you will be likely to cut will have a 1-inch diameter. The majority of the tubing used will be smaller. Commercial work can involve much larger copper tubing and require larger cutters.

Tubing cutters are primary tools for plumbers. With the amount of use they get and their importance to your job, buy the best in all of your tubing cutters.

Quick-acting tubing cutters

Quick-acting tubing cutters (Fig. 6.38) cut tubing the same way as standard roller cutters, but they are faster. These cutters have a quick-release button. The button allows the cutter to have its jaws opened by a touch of the button. Standard cutters require turning of the handle to open the jaws. The time saved with quick-acting tubing cutters is minimal, but it adds up over the course of the year.

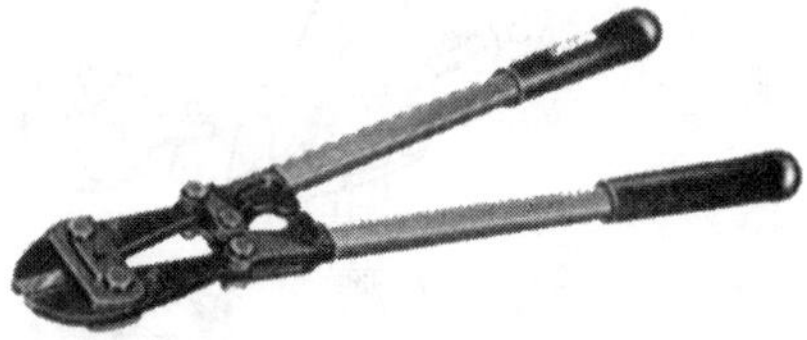

Figure 6.36 Bolt cutters. (*Ridge Tool Co.*)

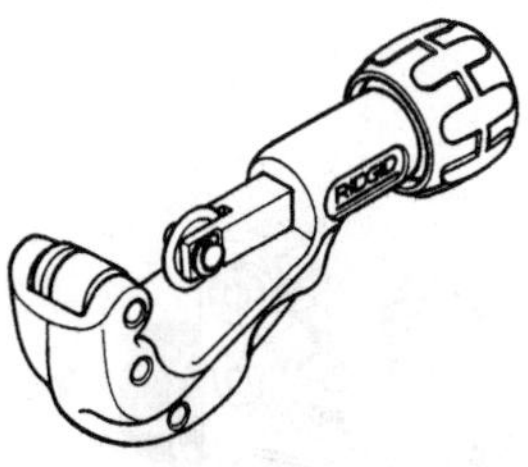

Figure 6.37 Standard tubing cutter. (*Ridge Tool Co.*)

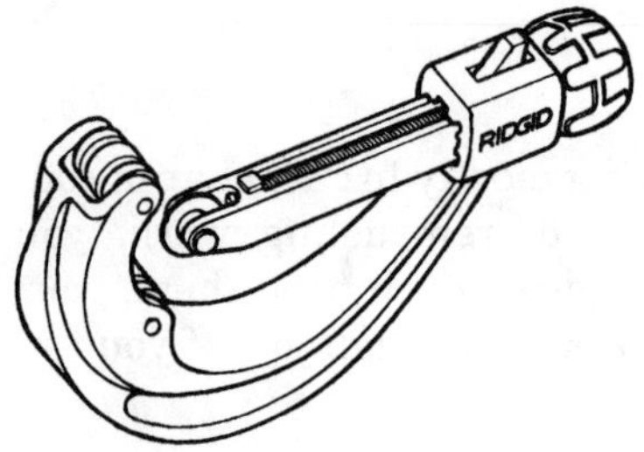

Figure 6.38 Quick-acting tubing cutter. (*Ridge Tool Co.*)

Midget tubing cutters

You have to buy a set of midget tubing cutters (Fig. 6.39). These cutters cut tubing just like other roller cutters, but they are very small and can operate in spaces where normal cutters cannot. Because of their diminutive size, it is difficult to get leverage on mini-cutters. For this reason, it is important to buy the best you can find and to keep them well lubricated.

Cutter wheels

The cutter wheels in tubing cutters get dull and worn out; sometimes they chip and break. Cutter wheels are not expensive, and you should keep a few extras on hand. Prices on these items are about the same across the board.

Internal tubing cutters

Internal tubing cutters (Fig. 6.40) are nice to have, but they are not used much. Most experienced plumbers will have a set, but apprentices don't need to invest in internal tubing cutters.

The cutters are used to cut pipe from the inside out. Their most frequent use is for cutting drain pipes that will connect to closet flanges. Since the pipe must be cut below floor level for the closet flange, there are times, as in the case of a concrete floor, when internal tubing cutters are about the only way to get the job done.

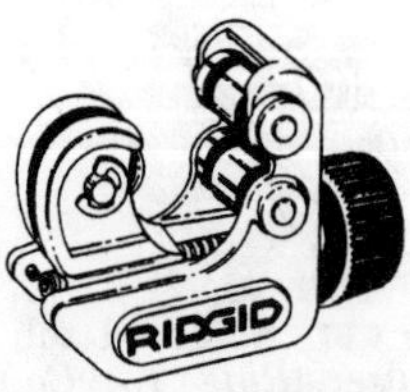

Figure 6.39 Midget tubing cutter. (*Ridge Tool Co.*)

Figure 6.40 Internal tubing cutter. (*Ridge Tool Co.*)

Ratchet-style soil pipe cutters

Ratchet-style soil pipe cutters (Fig. 6.41) are a near necessity for plumbers involved in remodeling work. These cutters allow cast-iron pipe to be cut without breaking a sweat. Because of their ratchet design, these cutters can be used where other cutters cannot. Make no mistake, ratchet cutters are expensive, but they are worth it. Fortunately for you, these tools are generally supplied by the employer.

Ratchet cutters cut cast iron with the use of numerous cutting wheels. The wheels are part of a chain that is wrapped around the pipe and secured to the cutter. The tension of the chain is set and the ratchet handle is pumped. After a couple pumps, the cast iron cuts smoothly and evenly. When you get to the point of wanting your own ratchet cutters, buy the best.

Scissor-type soil pipe cutters

Scissor-type soil pipe cutters (Fig. 6.42) use the same type of cutting wheels and chain used with ratchet cutters. But the operation of the tool is very different. These tools require a good deal of space to operate. The chain is secured around the pipe, and the cutter's handles are spread apart, like open scissors. Typically, one handle is resting on the ground or floor, and the other is extended upward. The worker applies

Figure 6.41 Ratchet-style soil pipe cutters. (*Ridge Tool Co.*)

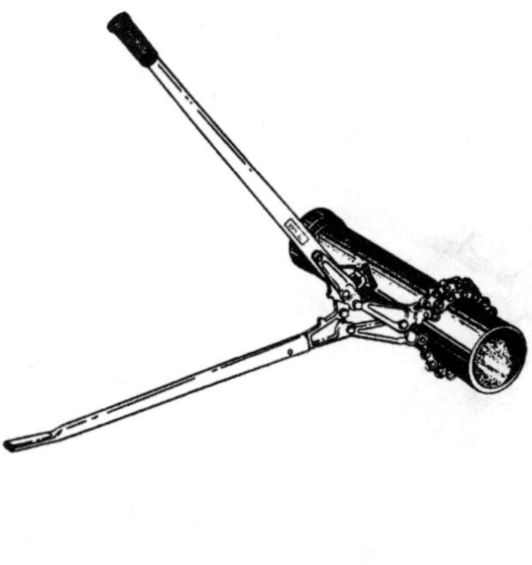

Figure 6.42 Scissor-type soil pipe cutters. (*Ridge Tool Co.*)

downward pressure on the elevated handle, and the cutting wheels snap the pipe. This type of cutter is often called a snap cutter.

Snap cutters work well when there is adequate room to use them. Like ratchet cutters, snap cutters are expensive and usually supplied by the employer. These tools are rarely used in modern plumbing, but they do find some use on commercial jobs. Quality is important in the selection of snap cutters.

Knockout Sets

Knockout sets (Fig. 6.43) are used to make holes in stainless-steel sinks. They are particularly useful when an air gap or accessory is being added to an existing sink. These tools are expensive, and they are not needed by apprentices.

Aviation Snips

Plumbers and apprentices use aviation snips (Fig. 6.44) to cut hanging straps for their pipes and for other odd jobs. You should have a pair of snips. Buy good ones. Cheap snips bend more metal than they cut.

When you look at these snips, you will find some variations in them. Some are made to cut straight, and others are made for cutting to the right or the left. Mechanics working with sheet metal carry all three, but a pair of straight-cutting snips will meet your plumbing requirements.

Figure 6.43 Knockout set. (*Ridge Tool Co.*)

Figure 6.44 Aviation snips. (*Ridge Tool Co.*)

Nail Pullers

All apprentices should have nail pullers on their lists of tools to acquire. There will be countless times throughout your career when a nail will have to be removed from your work area.

The Rest

The rest of the tools covered should be supplied by your employer. I mention them here only to make you aware of their existence and purpose.

Large sewer machines

Large sewer machines (Figs. 6.52 and 6.53) are used for cleaning drains with diameters of 3 inches or more. They can be equipped with a variety of heads for different types of stoppages.

Figure 6.52 Large sewer machine. (*Ridge Tool Co.*)

Figure 6.53 Super-sized sewer machine. (*Ridge Tool Co.*)

Small drain cleaning machines

Small drain cleaning machines (Figs. 6.54 through 6.58) are used for drains with diameters of 2 inches or less. Some models can be effective on drains as large as 4 inches. Assorted heads are available for various tasks.

Figure 6.54 Mid-size sewer machine. (*Ridge Tool Co.*)

Figure 6.55 Briefcase-style sewer machine. (*Ridge Tool Co.*)

Figure 6.56 Stand-mounted small sewer machine. (*Ridge Tool Co.*)

Figure 6.57 Cordless hand-held small drain cleaner. (*Ridge Tool Co.*)

Figure 6.58 Hand-held electric small drain cleaner. (*Ridge Tool Co.*)

Power threaders

Power-threaders (Fig. 6.59), used for threading pipe, make a hard job easier and faster. These units are very expensive, but if a company is doing a volume of threaded-pipe work, they are worth it.

Ratchet threaders

Ratchet threaders (Fig. 6.60) are used to thread pipe manually. These tools are much less expensive than power threaders, and they do the same job. However, they are much slower and require considerably more effort than power threaders.

Oilers

Oilers (Fig. 6.61) are used when threading steel pipe. The cutting dies of the threading device must be oiled to keep them from being damaged. Many plumbers use only an oil can for small jobs, but oilers are the only way to go for production threading.

Figure 6.59 Power-driven pipe threader. (*Ridge Tool Co.*)

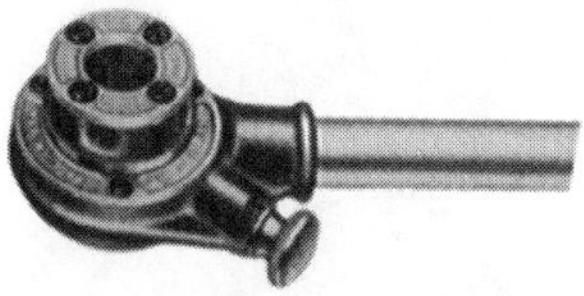

Figure 6.60 Ratchet-type pipe threader. (*Ridge Tool Co.*)

Figure 6.61 Oiler. (*Ridge Tool Co.*)

Tristand chain and yoke vises

Tristand chain (Fig. 6.62) and yoke (Fig. 6.63) vises are used to grip and hold pipe for cutting and threading.

Pipe reamers

Pipe reamers (Fig. 6.64) are used to ream burrs out of pipe that has been cut.

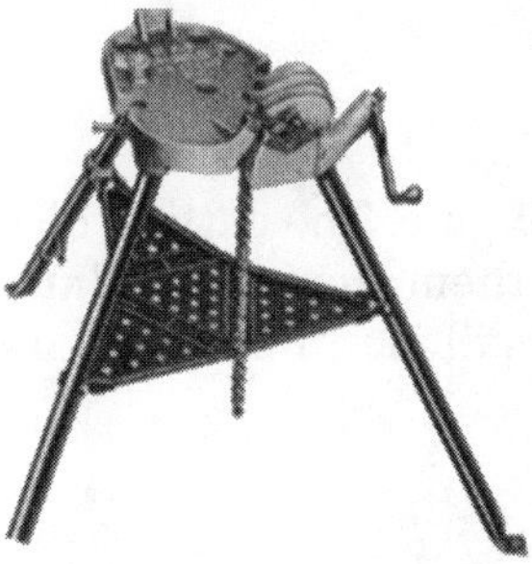

Figure 6.62 Tristand chain vise. (*Ridge Tool Co.*)

Figure 6.63 Tristand yoke vise. (*Ridge Tool Co.*)

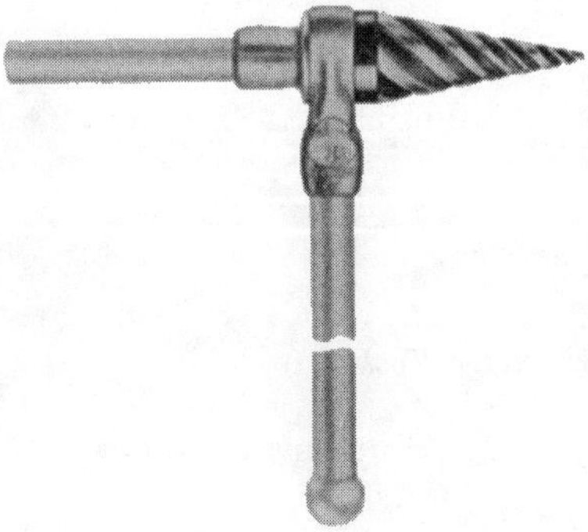

Figure 6.64 Pipe reamer. (*Ridge Tool Co.*)

No. 461 Straight Pipe Welding Vise — Used on mains and risers for aligning and holding the pipe which is to be tacked. Holds pipe from ½" through 8".

No. 462 Angle Pipe Welding Vise — Used on headers, drips and branches up through 12". Holds branch line at right angle to main or header for layout and cutting.

No. 463 Elbow Pipe Welding Vise — Used on mains and risers for lining up and holding 90° elbows at right angles to the pipe until tacked.

No. 464 Flange Pipe Welding Vise — Lines up face of flange at right angles to pipe and holds it there for tacking. Holds flange at any point below center of pipe.

Figure 6.65 Welding vises. (*Ridge Tool Co.*)

Welding vises

Welding vises (Fig. 6.65) are used more by pipefitters than by plumbers. They hold pipe in position while it is welded.

Four-wheel pipe cutters and hinged pipe cutters

Four-wheel pipe cutters (Fig. 6.66) are used to cut heavy metal pipe, like gas pipe. Hinged pipe cutters (Fig. 6.67) offer the advantage of doing this job in more restricted space.

Figure 6.66 Four-wheel pipe cutters. (*Ridge Tool Co.*)

Figure 6.67 Hinged pipe cutters. (*Ridge Tool Co.*)

Well, there you have it; we have covered all the tools you are likely to come into contact with as a plumbing apprentice. Our next subject will be blueprints; you will learn how to read and interpret blueprints.

Chapter

7

Reading Blueprints

Working plumbers need to be capable of reading blueprints. These drawings tell plumbers where to put their pipes, how big rooms are, what the slope of the land is, and more. Most code enforcement offices require blueprints and specifications to be submitted before plumbing permits are issued. Plumbing estimators use blueprints to develop take-offs and bid jobs. Reading the riser diagrams associated with blueprints shows plumbers how the plumbing system is designed for installation. To put it simply, blueprints play a significant role in the lives of plumbers.

There are usually two different opinions formed when apprentices find they will have to learn how to interpret blueprints and riser diagrams. One group of apprentices will assume that reading blueprints is no big deal. The second group will be intimidated by what seems a formidable task. These two common assumptions are both wrong. Blueprints are not difficult to deal with when you know what you are looking at. On the other hand, inexperienced eyes will not get far in making sense of complex blueprints.

A lot of apprentices think blueprints are made up of a few lines and some symbols. In a way, they are right, but most blueprints are much more complicated than that. Simple line drawings may have only a few lines that represent walls and a handful of symbols to show the locations of windows, doors, and fixtures. This type of line drawing is often used to show the floor plan of a home, but it is not the type of plan used to build a house.

Commercial jobs require blueprints that can consist of dozens of pages and stand a foot tall when rolled up. These plans have elevation plans, riser diagrams, heating layouts, electrical plans, cross-sections, front elevations, and a lot more. If you know how to read a set of blueprints for a commercial job, your layout work is done for you.

Residential jobs frequently have detailed blueprints for their construction, but the mechanical aspects of the job are not as well defined as they would be with a commercial job. The locations of plumbing fixtures will be shown, but there probably will not be a riser diagram or layout for the piping included with the plans. Before a plumbing permit can be obtained for the jobs, riser diagrams must be drawn. Guess who draws them? Plumbers. That's right, you have to learn more than just how to read blueprints; you must learn some drafting skills.

Residential service plumbers are about the only plumbers who don't deal with blueprints in their daily work. Commercial service plumbers use plans to track pipe locations for repair. Remodeling and construction plumbers use blueprints frequently. And to pass the exam for your plumber's license, you will need to be competent in drawing riser diagrams and reading blueprints. So even if you will never use blueprints after passing your test, you must learn the basics to get your license.

Reading blueprints is a little different than reading a book. A book is filled with words that the average person can understand. Blueprints are made up of different types of lines and symbols that the average person will not understand. It is not that the plans are difficult to understand; it is only that the reader must have a basic foundation of knowledge for what the various lines and symbols mean. In a way, learning to read blueprints is like learning a foreign language, except easier.

Lines

Blueprints are made up of a lot of lines. The lines vary in width, darkness, and construction. There are dotted, solid, dashed, dark, light, thick, and thin lines. Each of these lines has a different meaning. Once you understand what the various lines mean, you will be able to understand the lines on any blueprints.

Blueprints are made to exacting standards. To maintain consistency, the various lines on the drawings mean the same thing on every set of prints you read. To make life easier, most blueprints are provided with a legend. The legend shows examples of the lines and symbols used in the drawings, along with words to describe their meaning.

Object lines

Object lines are dark, thick lines that indicate an object such as a wall. These are the lines you will see used to indicate the foundation and walls of a building.

Center lines

Center lines indicate the center of an object. You will often find these lines used to show the center of a circular object like a well or spiral stairway. Center lines are thin and made with alternating long and short dashes.

Dimension lines

Dimension lines indicate the length of objects. These are thin, solid lines with arrowheads on each end. There is generally a number in the middle of dimension lines. The lines start with blunt ends at the number and extent in opposite directions to span the distance of the object. The number indicates what the distance between the two arrowhead ends is.

Extension lines

Extension lines are the thin, solid lines that provide a stopping point for the pointed ends of dimension lines.

Leader lines

Leader lines lead a reader's attention to notes on the blueprints. These thin, solid lines are dark and connect objects with pertinent notes located away from the object. For example, you might see a symbol within the object lines of a plan that indicates a water heater. From the water heater symbol, a leader line might direct you to notes that indicate the water heater will be a 52-gallon electric heater.

Hidden lines

Hidden lines represent surfaces that are hidden from sight. For example, hidden lines might be used to show the location of a recessed medicine cabinet in a wall cavity. Hidden lines are made with thick, short dashes.

Cutting plane lines

Cutting plane lines show sections that may be removed for examination. These lines are thick, usually the thickest on the drawing, and end with arrowheads. These lines are usually made of uniform dashes, but they are sometimes constructed with units of one long dash and two short dashes.

Phantom lines

Phantom lines are thin and dark. They are constructed with units of one long dash, followed by two short dashes. These lines indicate the scope of movement a part has. For example, phantom lines can be used to show the path a door will follow when it is opened.

Broken lines

Broken lines show parts of an object are broken away. They are thin, solid lines that are marked with a Z.

Contour lines

Contour lines show ground elevations. These lines are important when determining the depth requirements of sewers and water services.

The contour lines indicating existing grade conditions are usually made with thin lines. There will be a number written within the lines to indicate the existing grade.

Contour lines showing what the finished grade will be when the job is done will be drawn with dashed lines. Again, numbers written between the lines will indicate the new grade.

The degree of slope in an elevation is determined by the position of the contour lines on the plans. If the lines are drawn close to each other, they indicate a steep grade. Lines that are far apart indicate a rolling to level grade.

Cross-hatch lines

Cross-hatch lines are used to make a clear statement on blueprints. These lines are solid and very thin, the thinnest on the plans. Cross-hatch lines can show a cut surface, and they are used often to identify the difference between similar objects.

Now you know what types of lines you are going to be working with. The next issue of discussion will be the type of plans you will work with.

Plans

Just as there are many types of lines used to draw blueprints, there are also many types of plans associated with blueprints. These plans cover different phases of the job. Let's see what they are.

Working plans

Working plans are the blueprints used to construct a building. These are the master plans. They contain many smaller plans that assist in making the complete set of working plans. Not all elements of working plans will affect plumbers on every job, but a competent plumber will have the knowledge to understand all aspects of working plans.

Mechanical drawings

Mechanical drawings are the elements in working plans that indicate the placement and specification of mechanical systems such as plumbing, heating, and electrical systems.

Mechanical drawings are generally prepared by architects and drafting firms, but occasionally the trade contractors will draw them. These drawings show the locations and routes of plumbing, heating, cooling, and electrical systems and equipment.

The plumbing section of the mechanical drawings will generally be drawn in two different methods. One drawing will show the pipes and fixtures placed on the floor plan. Another drawing will be a riser diagram (Figs. 7.1 and 7.2), showing the plumbing system in a vertical design.

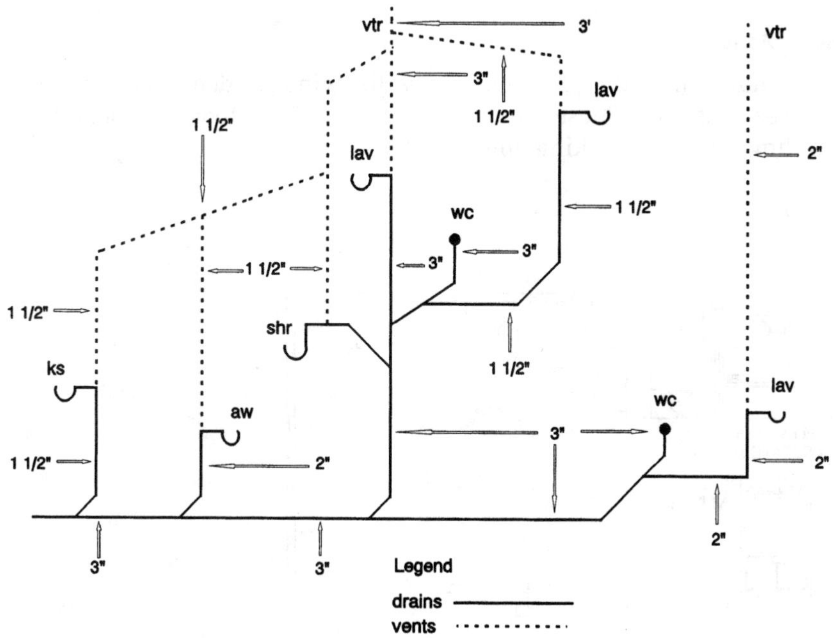

Figure 7.1 DWV riser diagram.

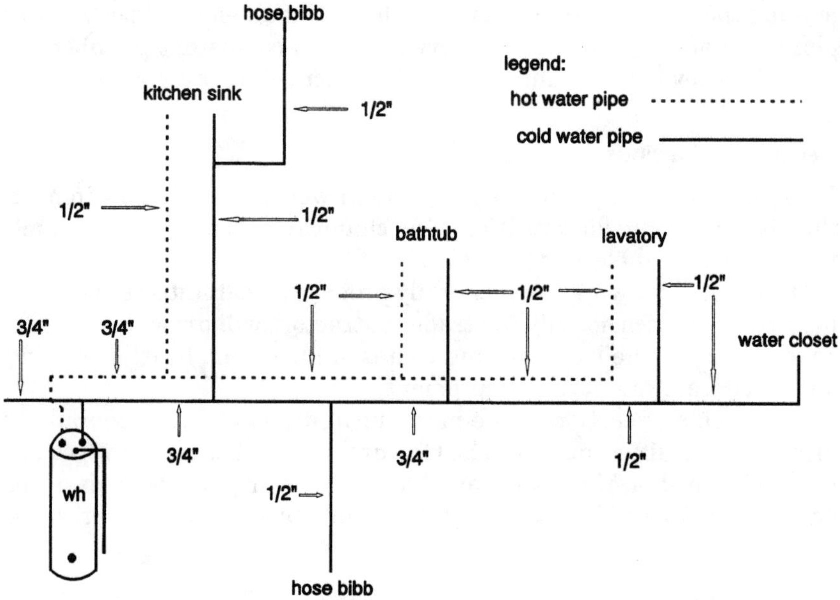

Figure 7.2 Water distribution riser diagram.

Floor plans

Floor plans show the placement of walls, window, door, cabinets, and fixtures (Fig. 7.3). These plans are drawn as if you took the roof off the building and were looking down on it.

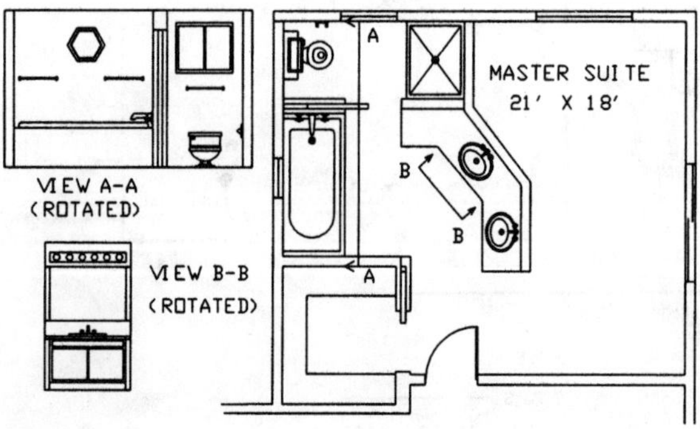

Figure 7.3 Floor plan.

Joist details

Joist details are drawn from the same perspective as floor plans. However, the joist plans don't show walls and fixtures. They concentrate on the placement of girders and joists. These plans allow plumbers to check for problems in routing their pipes through the structural members.

Cross-sections

Cross-sections are enlarged sections of the plans used to detail specific sections. For example, you might find a cross-section of the roof structure. This information will tell you if the roof will be framed with the use of trusses or rafters, and it will give details on the spacing and construction of the roof structure.

Building elevations

Looking at building elevations is like looking at pictures of the building. The view is as if you were outside of the building and looking at it from a normal perspective. There is usually an elevation for the front, back, and sides of the building.

Site plans

Site plans, also known as plot plans (Figs. 7.4 and 7.5) give an aerial view of the location of buildings and appurtenances on a property. The property boundaries will be indicated, and the placement of buildings and improvements will be shown. For example, looking at a plot plan for a rural home would reveal the boundary lines of the building lot, the foundation of the building, the septic tank, the distribution box, the septic field, the well, and the driveway.

Contour plans

Contour plans show how the grade of a property is configured. They are very useful in planning the installation of sewers, water services, building drains, and hose bibbs.

Landscape plans

Landscape plans show the placement of sidewalks, shrubbery, and related items. These plans can be useful to the thoughtful plumbers who don't want to put a hose bibb behind a foundation shrub or a cleanout under a sidewalk.

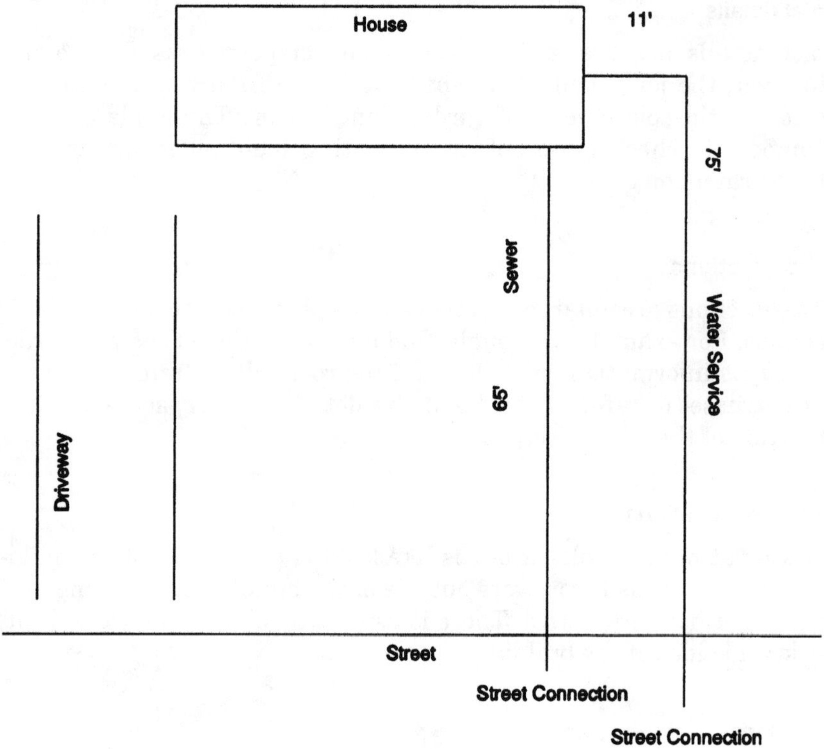

Figure 7.4 Site plan in urban location.

Scaled Drawings

Most blueprints are scaled drawings. When plans are drawn to scale, they are consistent in their measurements and can be converted to measurements that reflect their true size. Drawing life-sized blueprints would consume a lot of paper, and the plans would be very difficult to work with. Can you imagine unrolling 44 feet of paper to plumb a house? Well, if plans were drawn full-size, you could be in that position, and doing commercial work could require blueprints that unrolled for hundreds of feet. Since it is not practical to draw working plans in life-size dimensions, scales are used. The scales vary, but they allow large buildings to be drawn in realistic perspectives on small sheets of paper.

Most blueprints are drawn to a scale where every ¼ inch on the plans

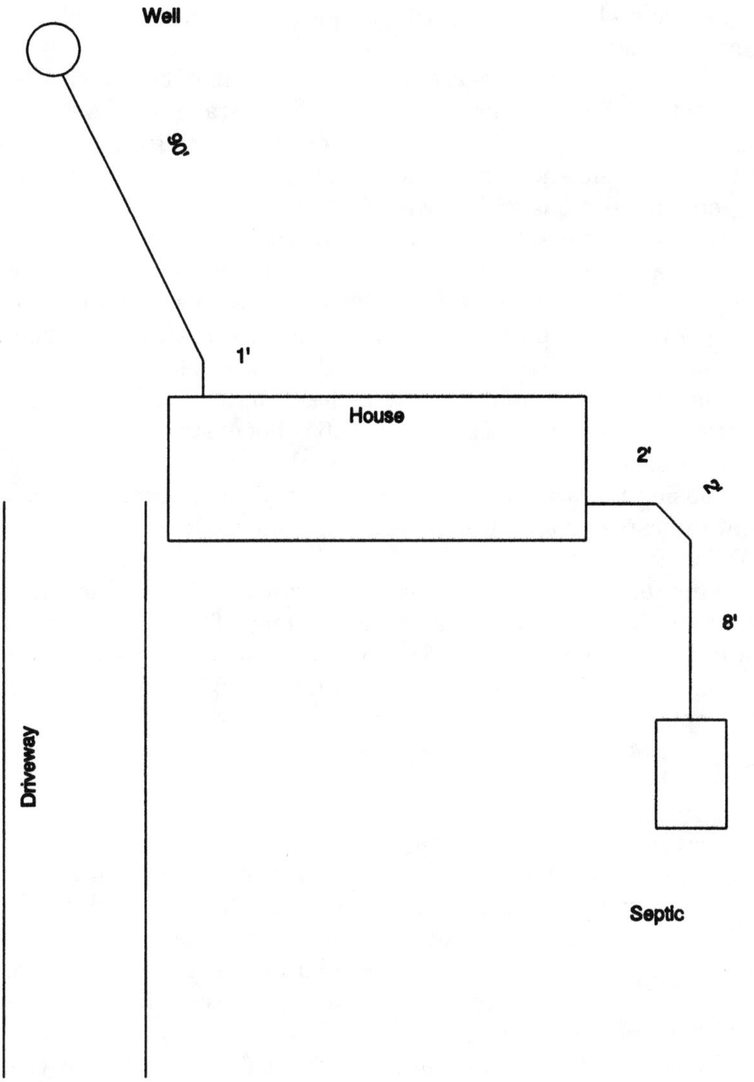

Figure 7.5 Site plan in rural location.

is equal to 1 foot in real life. For example, if a plan is using a ¼-inch scale, a 10-inch line represents 40 feet. By using a consistent scale in the preparation of plans, it is possible to capture the details of a building in proportion on manageable paper.

As I said, most plans are drawn with a ¼-inch scale. But some plans use different scales. Small buildings might be drawn with a ½-inch scale where a 10-inch line equals 20 feet. Big buildings might be scaled out using a scale of ⅛ inch. With these plans the same 10-inch line would represent 80 feet.

It is not uncommon for large sections of blueprints to be drawn with a ¼-inch scale and for accompanying details to be drawn with a ½-inch scale. This is something you have to watch out for. If you get lulled into using the ¼-inch scale and don't notice the notes that indicate a change in scale, your calculations will be wrong.

All plans will be labeled to show what scale they are using. If you are in doubt, you can check an item on the plans that you know the size of. For example, you know a standard bathtub is 5 feet long. If you measure the bathtub on the plan, you can figure out the scale of the plans. If the tub measures 1¼ inches, you are working with plans drawn to a ¼-inch scale. If the tub measures 2½ inches, you are dealing with a ½-inch scale. Of course, it is possible the tub is not of standard size, but this type of testing will give you the scale of the plans.

An even easier method is to measure the dimension lines. If you have a line that indicates an object is 40 feet long, you can measure the line and establish the scale.

Some plans, or parts of plans, are not drawn to scale. The plans should be marked with notes to indicate the lack of a scale, but don't count on it. If your work is dependent on the scale of the plans, check several items to confirm the consistency of the scale. Plumbers are not the only people who make mistakes; drafters can mess up too.

Scale Rules

You can scale plans with any type of tape measure or ruler, but a scale rule is the easiest way to measure scaled plans. There are different types of scale rulers; they include architect's, engineer's, and metric scales. For plumbing and working plans of buildings, architect's scales are the best. These scale rulers are marked with all the common scales used in drawing blueprints.

When you use a scale rule, the ruler does the scale conversion for you. Instead of using your tape measure and figuring out that 10 inches equals 40 feet, all you have to do is look at the scale ruler, and you will see the 40-foot measurement. I strongly recommend that you purchase an architect's scale ruler.

Plumbing Lines

Whether your are reading blueprints or drawing riser diagrams, you must be familiar with the plumbing lines used on the drawings. These are lines that represent different aspects of plumbing. Let me explain a little more about these lines.

Cold water pipes

Cold water pipes are represented on drawings with alternating long and short dashes.

Hot water

Hot water pipes are shown as a series of units consisting of a long dash and two short dashes.

Drain pipes

Drain pipes are indicated by the presence of a solid line.

Vent pipes

Vent pipes are drawn as a series of short dashes.

Gas pipes

A gas pipe is drawn with a line made up of long dashes where the letters FG separate the dashes. The FG stands for fuel gas.

Compressed air pipes

Compressed air pipes are drawn with moderate dashes that are separated with the letter A.

Fire protection pipes

Fire protection pipes are seen as a short dash, the letters FP, a long dash, the letters FP, and a short dash.

Vacuum pipes

Vacuum pipes are drawn with a series of dashes similar to those used for fire protection pipes, but the letters separating the dashes are VAC.

Hot water recirculating pipes

A line that is drawn with a moderate dash, three short dashes, and a moderate dash indicates a hot water recirculating pipe.

This covers the plumbing lines you are likely to encounter. Always check the legend on the plans you are working with to establish the meaning of various lines and symbols.

Symbols

Symbols are used to represent plumbing fixtures, equipment, and devices. There are symbols for bathtubs, toilets, sinks, water heaters, floor drains, hose bibbs, and any other plumbing part you might run into. There are even symbols to indicate the presence of gate valves and gas cocks. Figures. 7.6 through 7.23 include many sample symbols.

Manual Drafting

Manual drafting is fast becoming a thing of the past. Computers are being used more and more to replace the precision of a drafting pencil

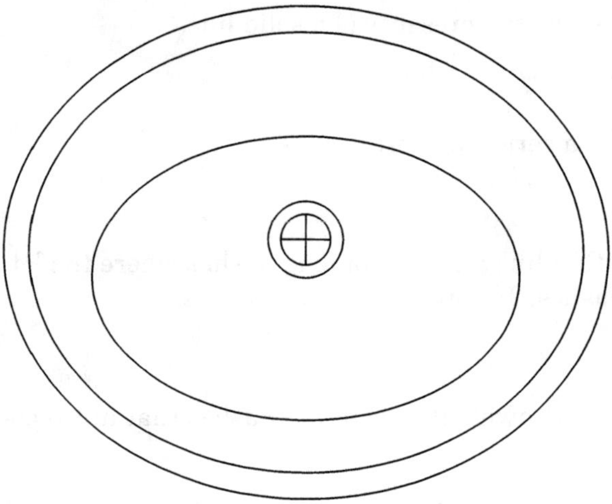

Figure 7.6 Lavatory.

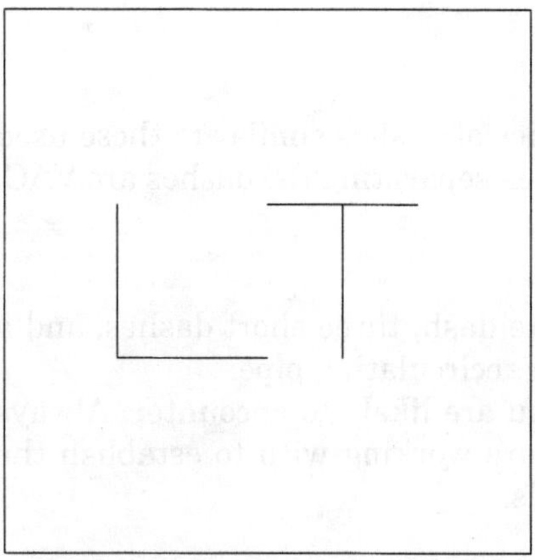

Figure 7.7 Laundry tub.

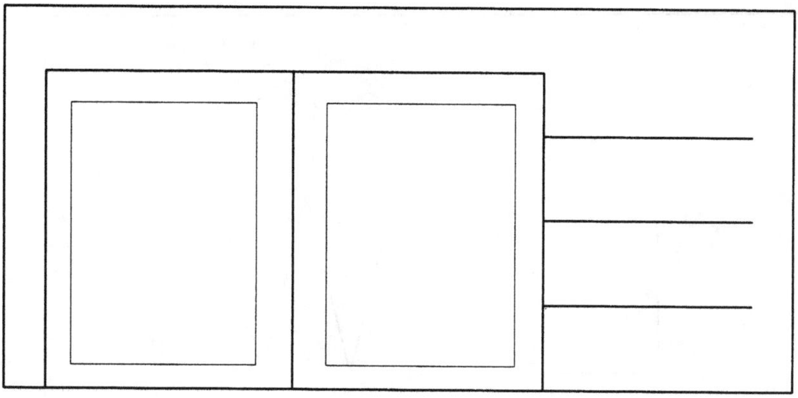

Figure 7.8 Double-bowl kitchen sink.

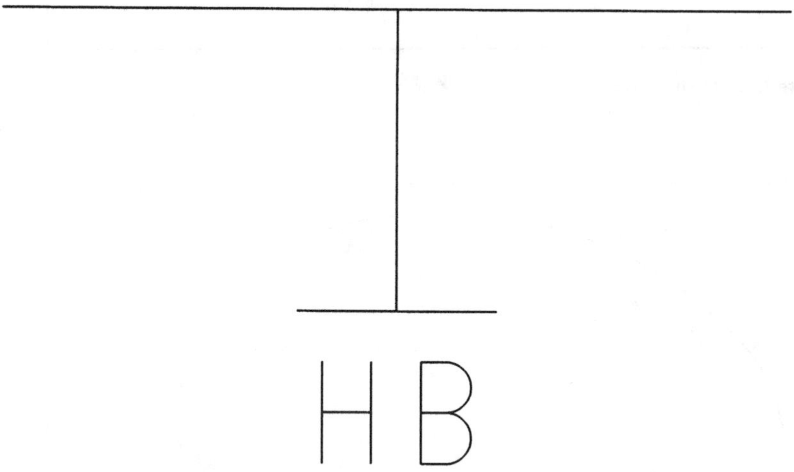

Figure 7.9 Hose bibb.

with the speed of technology. However, not everyone wants to use a computer to draw riser diagrams. And it is unlikely that you can take your laptop computer and portable printer or plotter with you on the day of your plumbing exam. So we are now going to talk about manual drafting.

Let's start with the tools used in drafting. Paper is the first need, and it should be graph paper: the paper with measured squares on it. Then you will need a #2 pencil. If you draw like most people, you will need an eraser. Finally, you will need a straight edge; plastic triangles that allow drawing angles and straight lines work well.

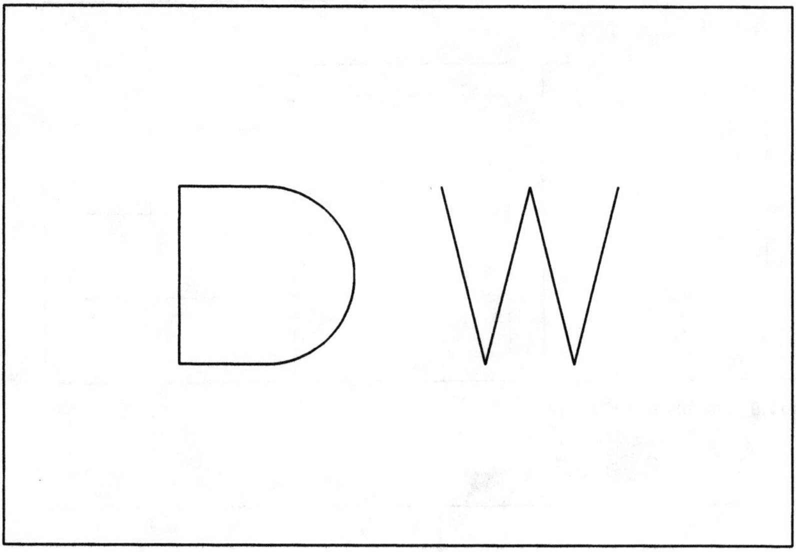

Figure 7.10 Dishwasher.

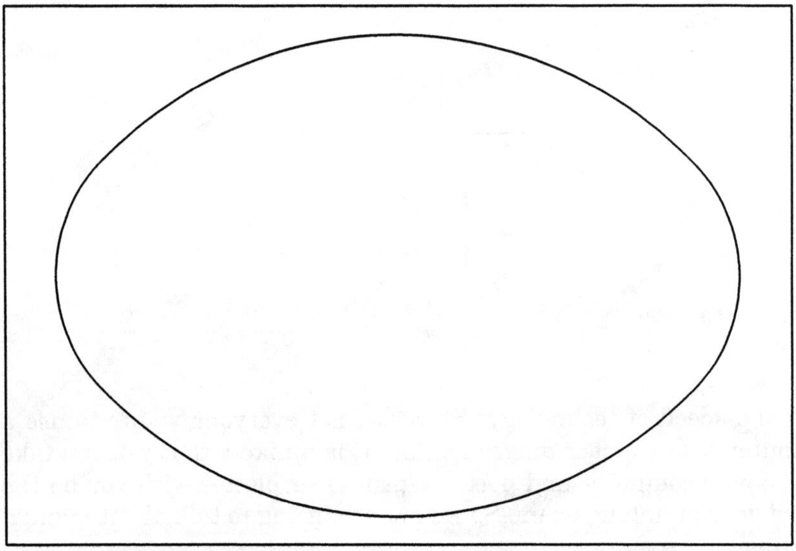

Figure 7.11 Whirlpool tub.

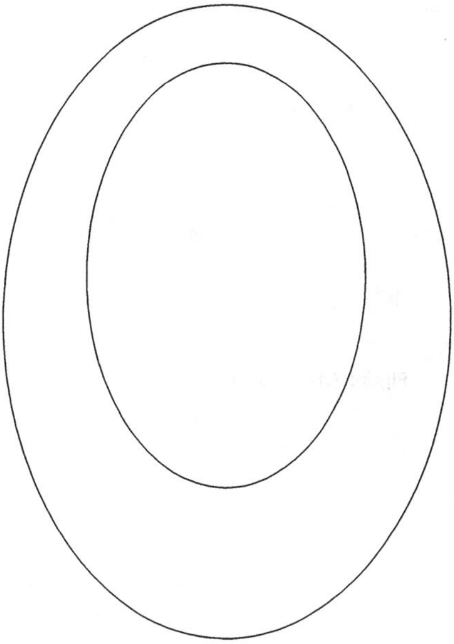

Figure 7.12 Bidet.

Figure 7.13 Shower.

You can enhance your drawing ability with a few additional tools. Choose a smooth, clean surface to work on. Your scale rule will help keep your drawings in uniform proportion. A drawing compass will allow you to make accurate circles. Various drafting templates will allow you to draw circles, letters, symbols, and so on. A book that teaches drafting techniques will be of great help.

Figure 7.14 Sump pit.

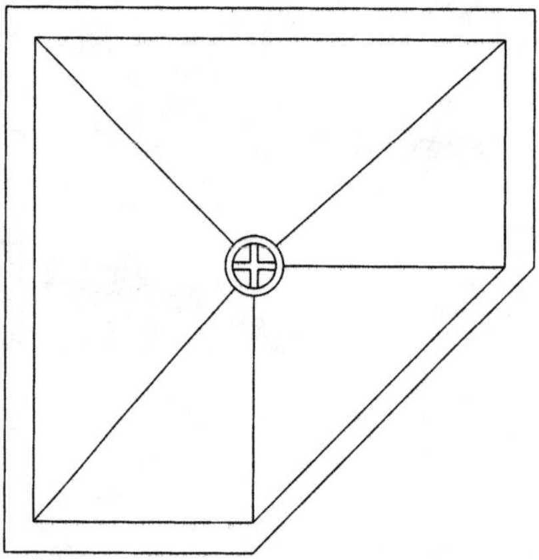

Figure 7.15 Corner shower.

Aside from a personal instructor, practice is the best teacher. Take your drawing tools and start drawing simple riser diagrams. The more you draw, the better you should get. As you become more comfortable with your drafting abilities, you can move on to larger projects. Stick with the riser diagrams until you master them; they are what you will draw the most of as a professional plumber.

Computer-Aided Drafting

Computer-aided drafting (CAD) programs are taking over the world. These high-tech programs allow people with limited artistic ability,

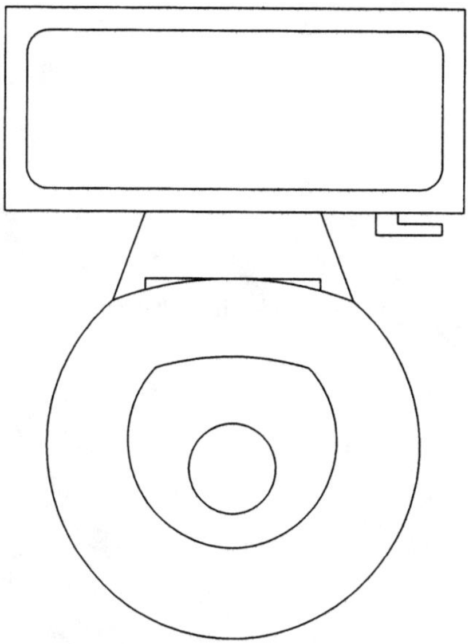

Figure 7.16 Water closet.

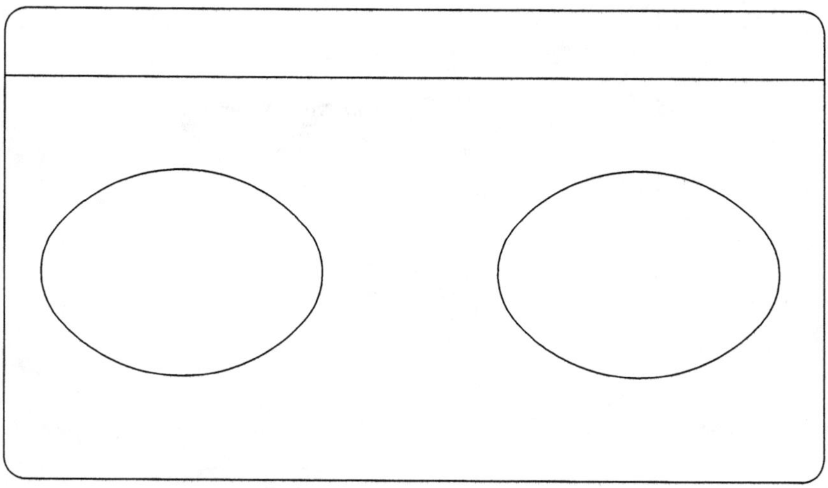

Figure 7.17 Double-bowl lavatory.

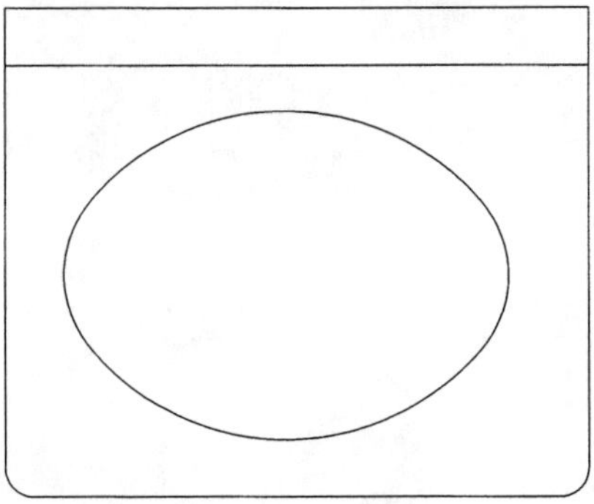

Figure 7.18 Molded lavatory top.

Figure 7.19 Water heater.

Figure 7.20 Floor drain.

Figure 7.21 Soaking tub.

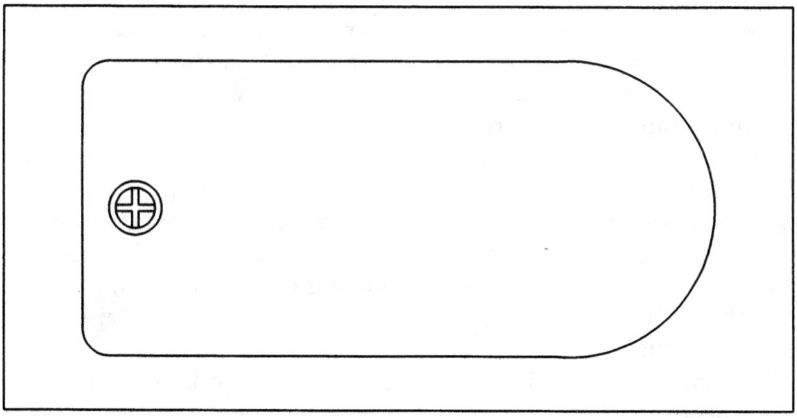

Figure 7.22 Bathtub.

Figure 7.23 Automatic clothes washer.

like me, to draw very well. The software programs often either include or can accept predrawn objects and symbols. These predrawn images speed up the drawing process and make it easier.

CAD programs take a while to get used to. Most of the software is fairly complicated. But once you know how to use it, the results of your on-screen drafting are impressive. Because of the complexity of CAD programs, I am not going to attempt to expound on them here. However, you should be aware of their existence.

Complex Drafting

Complex drafting skills can take years of concentrated study to develop. Since you are planning to be a plumber, you will not have to master complex drafting skills. When you get to the point where you can draw a clear riser diagram, you are well on your way to meeting your drafting needs.

While it is not necessary for you to harness the ability to draw complex blueprints, you must learn to read them. Let's spend the remainder of the chapter observing some examples of how reading blueprints will be important to your career.

To establish the importance of blueprint reading in the plumbing trade, let's take it from the top, or maybe I should say the bottom. Imagine yourself as a self-employed master plumber; feels good doesn't it? Okay, you have your own plumbing business, and a general contractor submits a set of plans for a housing development that will be built in the spring. It is your job to bid the job and provide specifications for the materials you will supply.

Before you can bid the job, you need to know what type of materials you will need and how many of them will be required. To determine this, you must do a take-off from the plans. If you can't comprehend the plans, you are in trouble.

In addition to your material requirements, you must estimate the amount of labor required to complete the job. Again, you must rely on your ability to read the prints to determine how much to charge for the job.

Once you have been awarded the job and start work, you will need to work with the plans again. This housing development contains houses with plumbing under concrete floors. The plumbing comes into the homes under the floors and turns up in walls to serve plumbing fixtures on the upper floors.

When you install your groundwork plumbing, you must be able to locate the walls your plumbing will turn up in. These walls don't exist yet, so you will have to use the blueprints to establish their locations.

If you miss the walls, you will have to break up new concrete and relocate your pipes; you don't want this and neither does the general contractor.

Once the houses are framed, you will have to rough-in the aboveground plumbing. You can't get away from those blueprints. While the prints probably won't show you where to put your plumbing, they will show you where not to put it. The prints will show you where windows and doors are going. You wouldn't want to run a plumbing stack up where a window or door is going to be located.

If the vanity will have a recessed medicine cabinet over it, the plans should show the cabinet. If you see it on the plans, you can offset your vent and avoid it. If you don't, you will have to cut and relocate your vent when the general contractor calls and complains.

The blueprints will also be used to establish the exact locations of the plumbing fixtures you are roughing-in. Make a mistake in scaling the distance to the center of the toilet you are roughing-in, and you will regret it. From start to finish, you will depend on blueprints to help with installing plumbing in the homes.

If you choose to tackle a commercial job, the blueprints will take on even more importance. Typically, commercial jobs must be plumbed exactly as they are shown on the plans. There is much less room for personal creativity in commercial jobs. You go by the plans, or you don't get paid.

The symbols for various valves, fixtures, and pipes will have to be second nature to you. If you spend an extreme amount of time pouring over the plans, profits will go down the drain. Plumbing is a competitive business, where production counts.

Whether you are a master plumber running your own business or an apprentice trying to get licensed, you must know how to read blueprints. The task is not difficult. All that is required is some background on what the lines and symbols mean. Once you know the principles, interpreting plans is easy.

A part of mastering blueprints is the ability to use math skills. The math is generally not complicated. However, there are some aspects of the trade where math is a little more difficult than adding or subtracting a few numbers. For example, computing the needs for roof drains can get a little complicated. In the next chapter we are going to look at these math factors.

Chapter

8

Math for Plumbers

The mathematic skills needed by plumbers can scare some people away from the trade. If they look at a course outline of the math requirements for apprenticeship classes, they may automatically assume they don't have the aptitude for it. The math descriptions can be intimidating, and the skills required of plumbers for mathematics can get complicated. However, much of the learning process can be eased with the proper presentation.

For a lot of people, the way the math is described has much to do with their interpretation of how difficult it will be to learn. For example, if I told you we were about to study linear measurements you might get tense. You shouldn't; linear measurements are the type used every day to measure how far you have driven or how tall you've become. They deal with the length or distance of a straight line. If you can read a tape measure, you know how to determine linear measurements. This part of the plumbing math is simple.

Do you know how to find the square footage of your home? It's not complicated, all you do is multiply the length by the width, and you know the square footage of the foundation. For example, if your home was 24 feet wide and 44 feet long, it would contain 1056 square feet. All that is needed is to multiply 24×44 for an answer of 1056.

When you delve into the math requirements for plumbing, you will work with whole numbers, fractions, decimals, angles, volume, liquid measurements, cubic measurements, and more. These math functions will enable you to make take-offs of the materials you will need and will allow you to calculate other plumbing requirements. One example would be sizing a roof drainage system.

Before a plumber can size drains for a roof, many math calculations must be made. A determination about the surface area of the roof must be made. Any elevated roof sections must be accounted for to allow for excess water hitting the obtrusion and running down on the roof. Conversions may be necessary to figure the average annual rainfall in your city. The volume of water that can be handled by roof receptors and drain pipes must be calculated. There is more, but we will get into that later.

If sizing roof drains seems like it will give you a problem, you are probably right. Many experienced plumbers fail their tests for master plumbing licenses because of roof-drain calculations. The math is demanding of your concentration, but it is not difficult when you have been properly shown how to perform the calculations.

This chapter is going to touch on all of the most common math requirements you will run into during your apprenticeship and beyond. I believe my experience as a classroom instructor will help you to understand the math requirements for plumbing. In going over the math, I will give you examples and make the lessons as simple as I can. As long as you pay attention, you should be able to grasp the skills needed to perform the calculations used in plumbing.

I am going to assume you have a basic understanding of simple mathematics. By simple mathematics, I mean adding, subtraction, multiplication, division, and fractions. These are all general math skills taught from elementary school into high school. If you don't understand the most basic math principles, consult a tutor or books that build basic math skills.

The math we are going to concentrate on in this chapter is well beyond the basics, but it doesn't have to be difficult to learn. We are going to work with angles, circles, square roots, and other measurements. All of these measurements are used in the plumbing trade.

Squares and Square Roots

We are going to dive right in with the use of squares and square roots. These are parts of math many plumbers despise. However, it can be quite simple if you look at it properly.

What is the square of the number 2? The correct answer is 4. The answer is found by multiplying the number 2 by itself. In other words, the formula would look like this: $2 \times 2 = 4$. Finding the square of a number is the same as finding the number's second power. When you see the number 2 with a very small number 2 placed slightly above and to the right of it (2^2), you are seeing the representation of the number 2 squared. The small 2 that is placed to the right of the number is called an exponent.

When people are asked to find the square of a number, many of them are perplexed. They have no idea how to perform the math function, but the act of finding the square for a given number is really quite simple, especially if you are using a calculator. Can you find the square of the number 5? Of course you can; all you have to do is multiply 5 by 5 and you will find that the square of 5 is 25. In reality, finding the square of a number is very simple. It sounds complicated, but it's not. All you have to do to find the square is multiply the number by itself.

Square roots are a little more tricky. However, if you have a handheld calculator, square roots are no problem. The least expensive calculators will perform functions to find the square roots of numbers. There are also charts available to give you the information needed on square roots.

The manual method of finding a square root could boggle your mind. If I were to try explaining the process in this text, I doubt that many people would understand it. The procedure is complicated and involves a lot of what plumbers would call "mumbo-jumbo." Square roots are a good example of where math skills for plumbing alienate otherwise good plumbers.

There are plenty of math books out there to show you the step-by-step process of finding the square root of a number. If you enjoy doing math the hard way, you can certainly find volumes of reading material on the subject. However, since I must assume you are more interested in being a licensed plumber than a math professor, I'm going to show you the real-world way of dealing with square roots in your plumbing activities.

There are tables available that show, in detailed listings, the square roots of various numbers. These tables will serve you well in your day-to-day activities. But since you will not be allowed to carry notes into your testing area when sitting for your licensing exam, let me instruct you on how to find square roots with your calculator, since calculators are generally allowed during testing.

I have two calculators in front of me. One of them cost about $5 and the other one cost nearly $200. Both of these machines calculate square roots with deadly accuracy. Surprisingly enough, my expensive calculator requires more keystrokes to find the square root of a number than the cheap calculator.

With my inexpensive calculator finding the square root of a number is very simple. I push the button or buttons to record the number I want the square root of. After doing this, I merely push the button on the calculator that is marked with the square-root symbol. Instantly, the square root of the number is displayed. I'm quite sure you would prefer to find your square roots this way rather than trudging through a long and complicated mathematical formula.

Plumbers are not required to stand up at a chalkboard and scribble exponents, divisors, quotients, dividends, and similar math symbols and functions out in long and complicated formulas. All they are required to do is know their business. You can know the math portion of your business very well with the least expensive calculator. Now, let's talk about cubes and cube roots.

Cubes and Cube Roots

Cubes and cube roots are no more difficult than square roots. Finding the solutions to problems dealing with cubes of numbers is not difficult. What is the cube of the number 2? Don't feel bad if you don't know; a lot of people are not sure of the procedures to use in cubing figures. However, don't allow yourself to be scared away; finding cubes is easy.

Let's examine how to find the cube of the number 2. All you have to do is multiply the number 2 by itself two times. Are you confused? Let me show you what the formula looks like. To find the cube of the number 2, your formula would look like this: $2 \times 2 \times 2 = 8$; 8 is the cube of 2.

Just as there are tables to assist you with square roots, so are there tables to help you identify cubes. Using a table for cube roots is the quickest and easiest way to obtain them.

Metric Measurements

Metric measurements continue to inch their way into our lives. The metric method is supposed to be easier to understand than the methods commonly employed in the United States, but that matter remains up for discussion. In theory, the metric system is easier, but try telling some seasoned plumber that you have a piece of pipe with a diameter of 76.2 millimeters and see what the response is. I doubt that the response would be printable in a respectable book.

The pipe with a 76.2-millimeter diameter is really a 3-inch pipe, in common plumber's terms. Metric measurements have tried to break into the workplace of the United States, but they have met strong opposition. In some ways, such as soft-drink bottles and drug deals, metric measurements have made their mark. We recognize a 1-liter bottle, and most individuals who watch television have heard of drug deals measured in grams.

Metric measurements generally make sense, but most Americans are afraid to give in to their logic. We are content with our gallons, feet, and inches. However, your code book should provide a table showing the conversions of various common measurements to metric measurements. There will be times when the metric numbers are the best ones to depend on.

Shapes and Measurements

When you get into math for the plumbing trade, you cannot avoid dealing with shapes and measurements. You will deal with many different shapes, shapes like triangles, quadrilaterals, rectangles, squares, trapezoids, polygons, circles, and others. For someone who doesn't have a college major in math, these circumstances can seem threatening. However, don't let them get to you; I'm going to show you how to manage the challenges.

Triangles

Triangles are objects bounded by three sides. Since the three sides may take on various lengths, therefore creating unique shapes, there are many types of triangles. If you enjoy technical reading, a hypotenuse triangle is equal to the square root of the sum of the squares of each leg. Does that make any sense to you? I doubt it; I not sure if I understand it fully. That's okay, because I'm going to put these fancy functions into words you can associate with.

The hypotenuse of a triangle is the side that is opposite of the right angle formed by the triangle. The two lines forming the right angle are known as legs. Finding the area of any triangle requires that you multiply the base measurement by one-half of the perpendicular height. For example, assume the base line is 10 feet long. The height from the base line to the peak of the triangle is 6 feet. To find the square footage of the triangle, you would multiply 10 by 3 to determine that the square footage of the triangle is 30 feet.

Quadrilaterals

Quadrilaterals are figures that have four lines. Examples of quadrilaterals are squares and rectangles. In addition to squares and rectangles, a quadrilateral can be a rhombus, rhomboid, trapezoid, right trapezium, isosceles trapezoid, or a trapezium.

Looking at the examples of quadrilaterals and wondering how to figure the area within the lines can seem difficult. But it is not really that tough. You need to be able to find the area of these shapes because you might come across a roof in one of these configurations. Let me show you how to do the math.

Finding the areas for squares and rectangles is very easy; you simply multiply the length by the width. When you multiply these two numbers, you arrive at the square footage of the shape.

The trapezium requires the use of a different method in finding the sum of its area. To do this, draw a line between two of the opposite angles. In doing so, you will create two triangles. Once you have broken the trapezium into two triangles, you can measure the area of each triangle, as described earlier. When you know the area for the individual triangles, add them together for the total area of the trapezium.

To find the area of a trapezoid, find the sum of the two parallel sides and then divide the sum in half. Take that number and multiply it by the perpendicular distance between the two parallel sides. The result with be the area contained within the trapezoid. Let me explain a little further.

Assume you are looking at a roof with the following measurements: one of the parallel sides is 60 feet long, one parallel side is 56 feet long, the side of the roof that forms a right angle with one of the parallel lines is 40 feet wide, and the other end of the roof is 44 feet wide. The first thing you would do is find the sum of the two parallel lines. Adding 60 and 56 gives you an answer of 116 feet. Divide the 116 in half for an answer of 58. Now, take the length of the perpendicular distance, the line forming the right angle, and multiply its length by 58. The formula would look like this: $40 \times 58 = 2320$. This tells you that the trapezoid roof has an area of 2320 feet.

In the case of parallelograms, you multiply the distance of the base by the perpendicular height. A parallelogram is a quadrilateral where the opposite sides are parallel.

Polygons

Polygons are shapes formed by multiple lines. They can contain many more than four sides. For example, an octagon has eight sides. Pentagons have five sides, hexagons have six sides, and decagons have ten sides.

To find the area of a standard polygon, begin by adding all the lengths of the lines. In other words, if you are working with a hexagon, you should have six numbers to add together. Multiply the sum of your six numbers by the perpendicular dropped from the center to one of the sides. Then, divide that number in half for the total area of the hexagon.

Circles

There are two numbers you must memorize and use when working with circles. The first number is 3.1416; this number is known as pi. You will use it in finding circumferences and diameters of circles. The second number needed in calculating circles is 0.7854. This number is used in finding the area of circles.

The circumference of a circle is found by multiplying the diameter by

3.1416. The diameter of a circle is found when the circumference is divided by 3.1416.

In order to find the area of a circle, multiply the square of the diameter by 0.7854. If you know the area of a circle and want to find the diameter, find the square root of the area and divide it by 0.7854. Finding the area of a sector of a circle can be done by multiplying the arc of the sector by half the radius.

Volume

It is often necessary for plumbers to determine the volume of an object or the volume capacity of a pipe or container. Finding the volume of a solid is accomplished by multiplying the area of the base by the perpendicular height. Arriving at the volume of a rectangular solid requires multiplying the length by the width and height. The surface of a cylinder can be figured by multiplying pi, 3.1416, by the diameter and then that number by the length.

Plumbing Situations

We could go on for chapters talking about volumes, trigonometric functions, and strange looking symbols and formulas. But this is a plumbing book, not a math book. Oh, I am well aware that you should have strong math skills to be a plumber, but I would prefer to teach you how to learn and apply the skills you will need in working with the plumbing code.

There are numerous books available to teach you hard-core math, and if you enroll in an apprenticeship class, your instructor will show you how to work the formulas needed in plumbing. The first part of this chapter has dealt with the theory of math; the remainder will deal with applying math in plumbing situations.

When you look through your code book, you will see many tables of numbers. These numbers and tables are used in working the math required for plumbing. Since most licensing exams allow the use of calculators and testing-facility-supplied tables, you can take some shortcuts on mastering math.

Apprentices gradually learn how to apply math to plumbing problems. There is, of course, the book learning, and that is reinforced with work in the field. Passing the exam for a journeyman's license will require some math, and the master's test will require even more math. However, the heavy-duty math in plumbing is usually done by engineers and architects, not plumbers. Let's look now at what might be found in the code that requires extended math skills.

Roof drains

Calculating roof drain sizes is the one area of plumbing math that seems to give plumbers the most trouble. You don't really need to know how to do this until you have advanced well into your plumbing career, but let me just discuss the basics of what is involved with figuring roof drains.

When figuring roof drains, you will typically be looking at a drawing that indicates roof dimensions and characteristics. You will need to be able to find the square footage of the roof area. We have already discussed how to find square footage, and doing so is pretty easy.

One piece of your sizing puzzle will be the amount of rainfall that can be expected to fall on the roof. The rainfall rate will be figured in inches per hour. In other words, you will need to know how many inches of rain will fall in a single hour. To determine the rainfall rate, you will be provided with a listing of states and cities that shows their rates of rainfall. As examples, Portland, Maine, has a rainfall rate of 2.4, and Richmond, Virginia, has a rainfall rate of 3.3. The rate of rainfall in Elko, Nevada, is 1.0.

As you can see, the rate of rainfall varies quite a bit from place to place. In figuring a roof-drain system for Richmond, Virginia, you would have to have a system capable of carrying more water than if the building was located in Elko, Nevada.

Since not all cities and towns are listed on the rainfall charts provided to you, it may be necessary to make conversions to arrive at local rainfall rates. This is done by either converting the drainage area or the tabular area of the tables provided for sizing. Does this sound complicated? Well, it's not.

If you want to convert the drainage area to the tabular area, all you have to do is multiply the drainage area by the rainfall rate. If you prefer converting the tabular area to the drainage area, you divide by the rainfall rate. This is only simple division and multiplication.

Other variables needed for sizing your roof-drainage system include the size and the slope of the drain. You will be provided with a table that gives options for different pipe sizes and grades. When you look across these tables, it is easy to match pipe sizes with surface areas. There will be a table of horizontal piping and another one for vertical piping.

Horizontal pipes will be rated for less square footage because they will not be running at full flow. Vertical pipes, on the other hand, can run at full flow and, therefore, are rated to carrying a higher volume of water.

So what extended math do you have to know to size a roof-drain system? None; you are working with tables and simple math. This is considered one of the most difficult math problems for the average plumber, and yet it only involves simple math. My point is don't get scared away from the trade by the math requirements; they are not that bad.

Sizing water pipes

Sizing water pipes can be quite complicated if you don't have prepared tables to work with. To size water pipes with the use of formulas, you must understand the friction head, the actual inside diameter of the pipe, the flow rate, and so on. This gets into some heavy-duty math. For example, you must account for friction loss, and that will depend on the type of pipe and fittings being used, among other things.

Fortunately, most plumbers never size water pipes by working with complicated formulas. Instead, they turn to a table in their code books and size their pipes in accordance with the requirements in the table. What kind of math is needed for this job? Not much. All you have to do is calculate the linear footage of the pipes being installed and refer to the table for sizing.

Waste and vent sizes

Waste and vent sizes could be determined with fancy formulas, but most plumbers rely on the tables provided in their code books. If your goal is to be a licensed, working plumber, learning how to use the charts and tables in your code book is all that will be required of you for sizing waste and vent pipes. If you are planning to become a plumbing engineer, you will have to learn the extended math and formulas.

Conversions

Conversions can require some knowledge of extended math, but the conversions you will need as a plumber should be listed in your code book. If they are not in your local code book, there are conversion tables available elsewhere for your use.

These conversion tables can show you how to make metric conversions and plumbing conversions. For example, your plumbing conversion table should show you that 1 gallon of water weighs 8.33 pounds. All of the conversion factors you should ever need as a plumber are available in these tables.

Other Tables

There are many other tables provided in plumbing code books. These tables allow you to quickly find the pounds per square inch for a head of water. You can compare atmospheric pressure and barometric readings at different altitudes. There are tables for assessing pipe expansion and degree day temperatures for cities across the country. In general, there is a table, chart, or illustration for just about all your math requirements.

Gas Pipe Calculations

A large number of plumbers work with gas pipe calculations. Since plumbers work with gas in many parts of the country, gas pipe requirements are generally a part of plumbing.

To determine gas pipe sizes with formulas, you must work with such factors as upstream and downstream pressure, viscosity, density, temperature, flow rates, and so on. Needless to say, this can be a complex task to undertake. Luckily, you don't have to size your gas pipe with complicated formulas; you can use the tables provided in your code book.

As you study your code book, you will find numerous tables for use in sizing gas pipe, if gas piping is a part of your local plumbing code. The tables will include such information as the length of the pipe being installed and the size of the pipe. There will be tables for different types of pipe and different gas pressures. All in all, these tables make sizing a gas pipe relatively easy.

So you see, the math requirements for real-world plumbing are not too difficult to handle. The key is having a calculator and the tables provided in your code book. When you learn how to use these two items, you can solve almost any math problem that arises in your job.

Chapter

9

Residential Plumbing

Residential plumbing is considerably different from commercial plumbing. Detailed piping diagrams are rarely available for residential jobs. It is normally up to the plumbing contractor to design the plumbing system to comply with local code requirements. For the plumber on the job, there might be many on-site changes needed in the initial plumbing plan. Since mechanical installations in residential work are not usually specified on the blueprints, the trades sometimes conflict with each other.

Residential versus Commercial Plumbing

You might say that residential plumbing requires more thought from a plumber than commercial work. In commercial jobs the plumbers have detailed piping diagrams to adhere to. With residential work the plumbers often have to make up the system as they go along. Since residential plumbers do often configure their own plumbing system, they have to hold a better knowledge of the plumbing code than some commercial plumbers do. This is not to say that commercial plumbers have no reason to know the code. It is just that commercial plumbers usually follow instructions and residential plumbers normally play it by ear.

I have done both types of plumbing, and I prefer residential work. There are pros and cons to each type of plumbing. Residential work is usually quieter and less structured than commercial work. Since plumbing a house doesn't take very long, residential plumbers get to change from job to job frequently. Commercial plumbers are often on the same job for months or even years.

The unorganized methods used to plumb a house can be enough to frustrate some plumbers. Unless the work being done is a housing project, where every fourth house is the same, residential work is rarely the same from house to house. The principles are the same, but the exact installations vary. Many plumbers don't like moving from job to job and always finding a new challenge. They prefer to stick to one large job and follow an organized plan for installing the plumbing.

There are also big differences between commercial and residential plumbing when it comes to repairs and service work. For example, replacing a residential garbage disposer is a simple job that one plumber can handle. Replacing a garbage disposer in the kitchen of a restaurant or hospital is a whole different story. These big grinding machines can require at least two plumbers just to move them. There are many similar situations that separate residential and commercial plumbing.

Do you have to choose between residential and commercial work? No, you can do both, but most plumbers settle into one or the other as their primary work. Some of your decision must be based on the work available in your area. It may be that when you enter the trade, the bulk of the work in progress is commercial work. This would probably cause you to choose commercial work over residential. If there is a healthy mix of both types of work, you can take your choice.

Just as most plumbers seek out a preference between commercial and residential work, so do most plumbing companies. It is certainly possible to find a company that does both, but it is common for companies to specialize in one or the other. There are many reasons for picking residential work as your primary source of income.

When people choose residential work over commercial work, it is not for money. Typically, commercial work pays better than residential work. As with anything in life, there are exceptions to the rule, but there is usually more money in commercial work.

If there is more money to be made in commercial work, why would anyone want to do residential work? I can't speak for other plumbers, but I can explain why I prefer residential plumbing. I enjoy the challenges of residential work. Trying to find a way to get a vent pipe from the basement to the roof doesn't frustrate me; I enjoy it. Residential jobs are not usually crowded with a lot of people, and I like the smaller working conditions. The variety of moving from job to job has always been an aspect of residential work that I enjoy.

General working conditions on residential jobs suit my preferences better. I prefer working with wood studs over metal ones, and I like wooden floor joists better than metal beams. The issue of safety can be confusing. I feel safer on residential jobs, but safety precautions are often better on commercial jobs. Since most commercial jobs are run to OSHA safety standards, they should be safer. Residential jobs rarely have the same stringent safety rules found on commercial jobs. It is rare to find a first-aid kit on a residential job site, trenches often don't meet minimum safety requirements, and wearing personal safety gear is not normally standard procedure on residential jobs. Even so, I still feel safer on residential jobs.

Commercial jobs are usually located in or close by cities. I don't like cities. I've always favored county living over city life. These feelings follow over into my work preference. Houses are built in the country, on the sides of mountains, and away from the hustle and bustle of cities. This allows residential plumbers, like me, to work away from the city.

As I said earlier, I have done all types of plumbing. I've plumbed muffler shops, schools, and health spas and worked on the plumbing in forensic labs. My work has included the installation of automatic pig waterers and $2500 gold faucets. The plumbing done throughout my career has been diverse, to say the least. Right now I want you to pull on your work boots and climb into my truck. I'm going to take you on a tour of what your job will be like as an apprentice with a residential plumber. Okay, get in the truck; I'll drive.

Working with a Residential Plumber

The job we are working on is a new house. The home will have an unfinished basement with plumbing in it and two floors of living space. There are two and a half bathrooms in the house. Our job today is to lay out and install the underground plumbing.

Now that we are at the job site, I'm going over to inspect the site while you unload the truck. Once you have the tools and materials near the work area, you will help me pull strings to indicate wall locations. We will drive stakes into the ground and string out all the walls that will affect our plumbing installation.

As the plumber, I have decided how the piping will be installed; there is no piping diagram on the blueprints. You give me a bag of flour, the kind people cook with, and I sprinkle it on the ground. The flour is used to indicate the locations and paths of the ditches that need to be dug. Guess what happens next? You pull on your gloves, if you're smart, and start digging the trenches. If the ground is soft or if we are working in a foundation hole filled with gravel, you can do most of your digging with a round-point shovel. If the ground is hard, you will have to use a pick to break up the earth. While you are digging, I'll be sorting fittings and putting the pipe and fittings near the ditches.

When you have dug enough of the ditches for me to start installing the plumbing, I will. If we were working on a larger project, one with multiple houses within easy walking distance, I might leave you to dig the trenches while I laid out another system.

When you have finished digging, you will help me by cutting pipe and delivering fittings to me. I might have you work on making a hole in the foundation or tunneling under it for the building drain to get outside of the foundation. If you pay attention, you will learn from watching me work. As the plumbing is installed, you will ease dirt back in the ditch to hold the pipes in place. But you must not cover the pipes completely; they have to be inspected by the plumbing inspector before they can be concealed.

Depending upon the site conditions, we might install the water service pipe underground. It will depend upon the anticipated final grading of the site. The water service might come through the foundation wall at a point much higher than the concrete floor.

When all of the plumbing is installed, we will use a tape measure to double check the pipe locations. We don't want these pipes that will be surrounded by concrete to turn up in the wrong places. When the installation is complete, you will assist me in capping the pipes and putting a pressure test on the system; we could use air or water to test the plumbing. When the test is complete, we will tidy up and prepare for the plumbing inspector. You will load the truck, and we will head back to the shop.

We will work on other jobs while we wait for the other trades, like the carpenters, to get the house framed up. When the roof is on the house, we will go back to do our rough-in work.

When we arrive on the job to rough-in the plumbing, you get to unload the truck. We will need pipe, fittings, tools, ladders, electrical cords, and other similar items. While you unload the truck, I will be checking the plumbing in the basement to make sure it wasn't damaged and didn't come up in the wrong spot. Sometimes concrete workers bump a pipe and knock it out of place.

After checking the groundwork plumbing, I will lay out the paths for the drains and vents to be installed. We start with these pipes because water pipes are easier to work with and can be offset around obstacles better than drains and vents can. While I am marking the locations of holes we will need drilled, you will be prepping the tools and sorting fittings into neat piles.

Once I have the holes marked, you will go to work drilling and cutting them out. While you are cutting the holes, I will be assessing water pipe locations and perhaps talking to the other trades to make sure our work won't be in their way. Then I will mark the locations for holes to run our water pipes through.

When all the holes are drilled and cut out, you will be my delivery person. I will yell out pipe measurements and you will cut and deliver the pipe. You will bring me fittings and occasionally help me with difficult phases of the installation. As I install the drains and vents, you will come along behind me and install pipe hangers. You might be sent to the roof to flash the pipes, but I may do it, depending upon your experience level.

I might have you cut some studs for use as backing to secure our pipes to. By the end of the day, you are going to be tired, but you still have to load all the tools and supplies back on the truck.

The next day we will install the bathing units. You will help me carry them into the bathrooms and set them in place. After we level the units, I will ask you to secure them to the studs.

Our next step will be installing the water pipe. If we are working with copper tubing and you have mastered soldering skills, I will fit the pipe together and you will solder the joints. I will cut the holes for the faucets in the bathing units, unless you are experienced enough to make the holes without damaging the bathing units.

As we move along with the water piping, you will install pipe hangers and kickplates: the metal plates used to protect pipes from nails and screws used by drywall installers.

When the rough-in is complete, you will cap all open pipes and prepare the system for testing. Once we have tested the pipes, using either air or water, we will make final preparations for the plumbing inspector. When we are done, you will load the truck, and we will return to the shop.

More time will pass as we wait for the house to near completion. In the meantime we will be working on other jobs. When the house is ready for us to set fixtures, you will be in charge of putting all the fixtures in the proper rooms. After unloading the truck, you will open the fixture boxes. I will probably have you start mounting faucets on the sinks and lavatories.

While you are preparing the fixtures, I will begin by installing the toilets and tub trim. When the lavatories are trimmed out, I will set them in place and connect them to the waste and water lines. If you're good, I'll let you install the kitchen sink.

Then we will go into the basement and install the water heater and laundry fittings. If we haven't already done it, we will connect the water service and sewer with their final destinations. When all the fixtures are set, we will test each one of them. After the job is cleaned up and ready for inspection, the plumbing inspector will do the final inspection and approve the plumbing. Our work will be complete, and we will go back to the shop and prepare for our next assignment.

Now that you have had a taste of what it's like to work on a construction job, let's see what you might do if you are assigned to a service and repair truck.

When you work as a residential service plumber, you often have to be on call to work nights and weekends, but the money is good for these overtime hours. Under normal conditions, you will work a typical 8-hour day and rotate being on call with other employees.

As the apprentice on a service van, you won't have to drill a lot of big holes or dig too many ditches. You will carry tools and parts in and out of houses and help the plumber in doing the repairs or service. Your first call might be a leaky faucet, and the second call could be a bathtub that won't drain properly.

Many service plumbers also do drain cleaning. This part of the job can get dirty and smelly. In general, you will ride from house to house fixing and maintaining all sorts of residential plumbing. You might work with well pumps, water conditioning equipment, and gas piping. The range of possibilities is extensive, but you won't be on most jobs for more than an hour or two.

There will be times when a sewer must be dug up and replaced, and you might have to dig up and replace a water service from time to time. If you are working in an area where freezing temperatures are present, you will spend a lot of time thawing and repairing water lines that have frozen.

Service and repair work rarely gets boring. Since it's conceivable to work in six to ten houses a day, you don't stay in one place long enough for boredom to set in. Most people consider service work to be much less demanding physically than construction work.

Residential Remodeling

Have you considered specializing in residential remodeling? Doing plumbing on remodeling jobs can be extremely challenging. Remodeling jobs rarely go as planned. Trying to get a 2-inch vent out of the roof of a home to vent a basement bath you are adding can require a lot of thought and planning. Even after you have a plan, existing conditions can alter your plans.

Remodeling work is my favorite form of residential plumbing. It is also one of the more profitable aspects of residential plumbing when you know what you are doing. Plumbing remodeling jobs can be very demanding on the mental capacities of plumbers.

If you decide to pursue remodeling work, you will have a lot more to learn than just plumbing. You will have to learn how to avoid cutting into hot electrical wires, how to deal with the bats, snakes, skunks, and other surprises you will find when opening old walls and crawling around under houses.

Creativity is an absolute necessity for plumbers working on remodeling jobs. You may have to figure out where to hide a vent pipe or how to get drains and water pipes up to a new attic addition. Remodeling work is full of surprises. If you venture into plumbing remodeling jobs, you are not likely to get bored.

More Residential Plumbing

Now you know a little about what to expect in residential plumbing. If you opt for residential work, you will learn much more on the jobs as your career advances. In the remainder of this chapter, I will give you more information on what to expect in residential work.

Most modern residential jobs are plumbed with plastic pipe for drains and vents. The water pipes in new residential jobs may be made of copper, CPVC, or polybutylene (PB). Most professional jobs are done with either copper or polybutylene. These lightweight materials make the work much easier than it used to be.

It was not long ago when drains and vents were made of cast iron and water pipes were made of galvanized steel. Both of these heavy materials are still available and approved for use, but they see little use in today's residential work.

The fixtures used in residential work are pretty consistent from house to house. You will work with kitchen sinks, bar sinks, laundry tubs, showers, bathtubs, lavatories, toilets, and water heaters on a regular basis. Some jobs will have whirlpool tubs and a few will have bidets. These are the most common fixtures found in residential jobs.

Gaining the skills to do residential work does not, in itself, qualify you for commercial jobs. The fixtures installed in commercial work can be vastly different from residential fixtures. You will learn more about those fixtures in the next chapter.

Plumbing is a trade in which you should never stop learning. Technology changes and brings new products to the market on a daily basis. If you become too complacent with your skill levels, you may find them outdated quickly.

How long will it take you to become qualified to plumb a house? If you have a strong desire to learn the trade and a plumber who is willing to work with you, it is possible that you could gather the basic skills to plumb a house in less than 6 months. However, many factors may stretch the learning process out to a full year.

The quicker you learn how to perform plumbing skills, the quicker you will get away from digging ditches and drilling holes. After learning to solder, you might be allowed to install water pipes while the plumber is installing waste and vent pipes. As your knowledge increases, another apprentice or a laborer might be used for the dirty work so that you can perfect your plumbing skills. The progression of your career is in many ways up to you and the effort you put forth.

Does residential service plumbing take long to learn? Yes and no. The plumbing code does not have a lot of affect on service and repairs. Therefore, once you have passed your licensing exam, you don't have to retain as much of the code as you would doing construction or remodeling work. However, the diversity of service and repair work makes learning this phase of the trade a time-consuming project.

In the case of construction plumbing, once you have learned the basics of piping and installing new fixtures, you can handle most of the job on your own. With service and repair plumbing, there are many type of fixtures and parts to work on, and working conditions vary on each job.

The fact that you know how to solder doesn't necessarily prepare you for repairing pipes that have frozen and split. Being able to replace a floor-mounted toilet doesn't qualify you to install a wall-hung toilet. The simple act of replacing a faucet can get complicated if the mounting nuts have rusted.

Service and repair work is something you learn as you go about your daily activities. Each job can present you with different circumstances that may call for creative thinking. Experience is the best teacher when dealing with the unknown variables you will find in plumbing work.

The mix of job opportunities between residential and commercial work will vary. Economic conditions play a part in which side of the trade offers the most work and so do geographic locations. The type of plumbing you do within either category of plumbing will also have an effect on how much work is available. Yet another factor will be the competition of other plumbers in your area. You will have to assess the area you will be working in to determine where the most job opportunities are.

We have pretty well covered the overview of residential plumbing. Let's move on to Chapter 10 and learn about commercial work.

Chapter

10

Commercial Plumbing

Commercial plumbing is only a distant cousin to residential plumbing. While many of the same pipes, fittings, and fixtures may be used, there are numerous materials used in commercial work that never find their way onto a residential job. There are not many houses equipped with bedpan washers or urinals. Plumbers who are very skilled in residential work can be lost on a commercial job. Conversely, commercial plumbers, who are used to having their jobs laid out for them, may be driven to the brink of madness trying to plumb a house or residential remodeling job. If you ask around, you will find little argument that residential and commercial plumbing are worlds apart.

The last chapter talked about the benefits of residential plumbing. This chapter is going to do the same for commercial plumbing. You will be introduced to the materials used on commercial jobs, and you will get an idea of what it is like to work the big jobs. Let's start this introduction and comparison with some of the various materials that separate commercial work from residential plumbing.

Materials and Fixtures

There are many materials and fixtures used in commercial and institutional plumbing that you will not find on a residential job. For example, how many houses are equipped with sterilizers, clinical sinks, or fluid suction systems? Well, I don't know of any, but hospitals are one place you can find these items. Do houses have roof drain systems? Not unless you count gutters, and they are not really the same thing. If you plumb a dental office, you will run into a lot of equipment to plumb that will never show up in the average home.

Let's move away from institutional work and look at an automotive garage. Not many houses have compressed air piping, but garages do.

Auto shops with grease racks, where oil and flammable materials may be produced, are required to have special separators. These separators separate the oil, grease, and flammable materials before the waste is allowed into the sanitary drainage system.

Commercial kitchens are equipped with grease interceptors. About the only grease interceptor you will find in a residential kitchen is the basket strainer in the sink. The water discharging from a commercial dishwasher can be hot enough to make the use of plastic drains near the appliance inappropriate. Residential dishwashers discharge through a rubber hose.

Pipe sizes on commercial jobs can be much larger than any you would work with in residential plumbing. Six-inch drains are common in commercial work, and larger ones are often used. The largest drain found on most residential jobs has a 4-inch diameter.

Have you ever seen a house with emergency shower facilities or an eye-wash station? Probably not, but you will find these fixtures on some commercial jobs. Wall-hung toilets can be found in either type of plumbing, but they are much more prolific in commercial work. Urinals are not normally installed in homes, but they are found in a lot of commercial buildings. The toilets in homes are generally flushed with the help of a flush-tank. Commercial toilets generally depend on a flush-valve, rather than a tank, for flushing. The shape of the toilets used in the two types of plumbing is another difference. Residential toilets are usually round-front toilets; commercial toilets generally have elongated bowls.

Water fountains and coolers are commercial fixtures that most homes don't have. Few residences will have any need for waste piping approved for handling chemical discharges, but you can find these needs in commercial jobs.

I've never seen a need for steel pipe with welded joints in residential work, but it is used on commercial jobs. Water heaters on commercial jobs are usually considerably different from residential water heaters; they are usually much smaller or much larger. In the case of a commercial job where only one bathroom is installed, the water heater will probably be very small and tucked under a cabinet or installed above the bathroom. However, water heaters for nursing homes and schools will be much larger than residential water heaters, and often there will be more than one.

Backflow preventers on commercial jobs can be very big compared to similar residential devices. Garbage-can washers show up on some commercial jobs but not on residential ones. In general, there are many differences in the materials and fixtures you will be working with on commercial jobs.

Blueprints

The blueprints used for commercial jobs are much more detailed than those used for building the average house. As a commercial plumber, you will have to adhere strictly to the plumbing diagrams on the plans. All of your work will be laid out for you. Unlike residential plumbing, you are not responsible for determining the most functional and least expensive ways of installing the plumbing. In fact, you have little to no control over how the plumbing will be routed. You must follow the instructions on the blueprints.

Blueprint symbols for the plumbing installation will be numerous, and you must understand what the symbols mean. These symbols will show the direction of the piping, the locations of various types of valves, fixture locations, and a lot more. Residential blueprints rarely include more than the symbols indicating fixture locations. Commercial blueprints show you everything about the installation requirements.

Safety Requirements

The safety requirements for commercial work are also more extensive than they are for residential work. Good safety practices should be used on both types of jobs, but they will be enforced strictly on commercial jobs. Failing to wear your hard hat could result in a significant cash fine. Not wearing safety glasses could also get you disciplined or fined. Before you set foot on a commercial job site, you had better know the safety requirements and follow them.

Working Conditions

The working conditions on a commercial job may not be anything like those of a residential job. The materials and fixtures used in commercial work differ from residential jobs, and so do the methods of installation.

Some commercial jobs are very large. You could spend days, possibly weeks, just installing floor drains. Typically, the ceilings in commercial buildings are much higher than those in houses. You will be working with metal studs, steel beams, concrete, and other materials regularly on commercial work. Extension ladders and high scaffolding will be used extensively, and you will often be surrounded by noise, moving equipment, and a horde of people.

While you might never install a length of cast-iron pipe in the plumbing of a new house, you could pipe the entire underground plumbing of a commercial building with cast iron. If you are running residential gas pipe, the largest size you are likely to use is 1-inch, and occasionally 1¼-inch. However, 2-inch pipe is common on commercial jobs, and you might get into larger pipe. Welding and working with mechanical joints will also be a new experience for the apprentice transferring from residential work to commercial jobs.

Instead of nailing lightweight pipe hangers to wooden floor joists for a residential installation, you may find yourself on scaffolding installing bar-joist hangers all day. These heavy hangers and the turning of the C-clamps can make your arms feel numb by lunch time.

It is not unusual for a single apprentice to serve multiple plumbers on commercial jobs. You could be stationed at a power threader, spending your entire day cutting and threading pipe. The monotony of this job can get to you quickly. If you are unfortunate enough to be threading the pipe by hand, like I used to, you will develop a strong upper body, if you survive it.

In many ways, the physical work involved with commercial plumbing is harder than that of residential work. There is a lot of climbing, heavier pipe, and more ditches to dig. The physical demands are more noticeable as an apprentice or helper than they are as a plumber.

Is commercial plumbing all doom and gloom? No; it offers many advantages over residential plumbing. Commercial jobs have the plumbing laid out for you. Unlike residential work, where you have to fight to get the plumbing installed sometimes, most commercial jobs go together like well-made machined parts. There will be some occasions when the job will not go as planned, but they are few and far between. If you don't like frustration and mental challenges, commercial work might be just the ticket for you.

Commercial jobs are usually well equipped with safety gear; this is another advantage over residential jobs. If you cut yourself on a commercial job site, a first-aid kit will be close by. All plumbing trucks should be equipped with first-aid gear, but they are not. Cutting yourself on a residential job could mean sticking a rag over the cut and securing it with duct tape. You shouldn't have any problem finding suitable first-aid materials on commercial jobs.

Since commercial jobs are often crowded with people, there is generally someone close by if you need emergency help. As an apprentice, you will almost never be left on any job alone, but as a licensed plumber, you might be the only soul on a residential job. Falling off a roof or ladder and breaking your leg, or worse, can be even more serious if you are working alone. There is safety in numbers.

While I say there is safety in numbers, let me make one point clear. It is a good idea to avoid working on jobs alone, but working in a crowd can increase the danger you are susceptible to. When you work by yourself, you only have to be concerned with your own actions. When you are working as part of a large team, you have to keep an eye open for what the other players are doing.

Storage trailers are usually provided for pipe and tool storage on commercial jobs. Being an apprentice, these trailers make your life easier; you don't have to load and unload the truck every day. You will, of course, have to deliver the goods from the trailer, but you can take it out in small quantities and avoid much of the restocking at the end of the day.

Commercial work offers the advantage of a more predictable schedule. An experienced plumber, with a good helper, can rough-in the plumbing for a house in less than a week. When that job is done, it may be questionable where the next job will come from. Even if work is lined up in advance, production schedules on residential work are rarely met. There could be days that you will not have a job to go to. This is not such a problem with commercial work.

Once you are working on a commercial job, you can get a good feel for how steady the work will be. If the job is a big one, like a health spa, hospital, or school, you are going to be there for a long time. The length of the job will be measured in months or years instead of weeks. This added security can help keep your bank balance healthy.

Speaking of bank balances, let's talk about the difference in pay scales for residential and commercial work. On average, commercial jobs pay better than residential ones. You probably won't see a lot of difference in your paycheck as an apprentice, but when you are a full-fledged plumber, you could.

How much more will you make as a commercial plumber? That is hard to predict, but you could make up to a few dollars an hour more than a residential plumber. Some commercial jobs are referred to as scale jobs. These jobs have posted notices that indicate the wages being paid to the various trades. During my career, I have seen a lot of different pay scales.

When I was working in Northern Virginia, I would talk to plumbers who worked in the District of Columbia. These plumbers were doing commercial work and making a lot more money than I was. I was doing residential work in Virginia. I was happy with what I was doing, but the increased wages tempted me to try commercial work in the District.

My past experience with commercial work had been in Virginia, and while I didn't like it, I could do it. I quit my residential job and went to work on a big commercial job in downtown D.C. My pay jumped by something like $1.50 an hour. I worked the job for less than a month when I realized I hated it. The extra money couldn't make up for the working conditions, the driving, and the surroundings. I quit the job and went back to work in Virginia, doing residential jobs.

There were many times when I would see help-wanted ads quoting pay scales well above what I was making. The spread between residential and commercial work ranged from very little to several dollars per hour. I did more commercial work as time went on, but I always came back to residential.

It is very possible to make an extra $80 a week as a commercial plumber, maybe more. It is also possible that you will not make any more than a residential plumber. It all depends on the job market and the economic conditions.

If you follow your plumbing career into owning your own business, you can make considerably more money on commercial jobs. The mere size of the jobs offers the opportunity for higher profits. However, beware; the size of the jobs also offers the business owner the opportunity for larger losses if mistakes are made in figuring the job.

I don't know if there is any fair way to say which side of the trade is the most profitable. Doing what you enjoy is worth a lot, and I guess I would recommend that you do the type of work you like and strive to make as much as you can doing it.

Commercial Service Work

Commercial service work is not the same as residential service. As you already know, there are many types of fixtures and equipment to work on with commercial buildings. The manner is which the work is done can also be very different. Let's look at when commercial service work is done.

Working hours

The working hours for commercial service work can vary a great deal. There will be many times when you will be asked to work in the middle of the night. Let me give you an example. Imagine that you are a commercial service plumber. Your employer tells you that tomorrow you will need to start work at midnight. Your job will be to repair some leaks in the hot water piping of a motel. Why would your boss want you to do the job at midnight?

The timing of the repairs would be scheduled in this way to avoid inconveniencing the motel guests. Since the job will require cutting off the hot water, the guests will not have any hot water while you make the repairs. If you did this work during normal waking hours, the guests would not be happy. By doing the job while the guests sleep, very few people are affected by your work. This type of middle-of-the-night work could apply to food stores, hospitals, nursing homes, and all sorts of other commercial buildings.

Work differences

There are also many work differences between residential and commercial service work. Just the size of the pipes can make a difference in your work. I talked earlier in the book about the size of commercial garbage disposers, but it is worth repeating. As a residential service plumber, you would have no trouble working on or replacing a garbage disposer. But when you tackle the same job on a commercial level, the work may require a second set of hands. Commercial disposers can be huge.

As a commercial plumber you will have to be familiar with grease interceptors, or grease traps as they are often called. You will work with the drainage of ice machines, food coolers, and rain leaders.

Snaking a drain in a home usually doesn't take very long, but the same type of job in a commercial building can be quite complicated. The long runs of pipe may mean you will have to move from cleanout to cleanout to clear the stoppage.

In general, service work in commercial buildings is similar to residential service only in its principles. You will work with a wide variety of fixtures, appliances, and materials that will never be found in a house.

Which Way Do I Go?

Apprentices are often confused about which field of plumbing to enter and ask themselves, "Which way do I go; which way do I go?" All apprentices have to make that decision on their own. No one can tell you which type of plumbing will suit your nature. The most effective way to make a sound decision is to try both types of plumbing.

If you will be working in a small town, you will probably have to switch back and forth from residential to commercial. Areas with small populations often cannot support specialized plumbers. Plumbers in these communities must do a little bit of everything to pay their bills.

You might find you enjoy the change of pace in switching between residential and commercial on a routine basis. In the early development of your career, getting experience in both fields of plumbing is a good idea. You will be a more well-rounded plumber in the long run. After you have learned to do all types of plumbing, you can decide if you want to settle into one particular field.

As an apprentice, you should concentrate your efforts on learning the trade, not finding a niche to cut corners on your learning process. You can, of course, get by with learning less when you stick to one style of plumbing, but you limit your career options.

You will have to learn the rules for all types of plumbing to pass your licensing exam, but learning the rules is not the same as learning the trade. Book-learning is good and needed, but it is not enough. You cannot consider yourself a competent plumber until you have done the work. Saying this reminds me of a man I would like to tell you about.

I worked for a company once where the owner and master plumber was a very good businessman. He knew how to get jobs and run a business, but he had very little mechanical ability. In those days, you could get a master's license by passing the plumbing exam. Requirements for classroom training and field training were not what they are today.

That businessman was very smart and had no trouble learning the plumbing code. If I had a question about the code, he could answer it without ever opening the code book. However, if I had a technical question about a problem in the field, he was stymied. This brilliant businessman was a master plumber, but he had almost no idea how to solve field problems. Why? Because he had spent little to no time in the field.

The man was smart enough to have a supervisor who could answer any question that came along in the field. But, what would the owner do if he were not surrounded by good field plumbers? My point is this: Learning the plumbing code does not make you a plumber. Holding a license that says you are a plumber doesn't make you a good plumber. The only way to become a good plumber is to combine classroom-type education with field experience.

Are you ready to move on to the meat of this book and learn how to install plumbing systems? Good, then let's do it. Chapter 11 is going to show you what you need to know about well systems.

Chapter

11

Well Systems

Well systems are the heart of rural plumbing installations. Without a well system, houses in the country have no water. These systems can consist of little more than a few pieces of pipe and a hand pump. However, for most homes, a well system is made up of many parts. There is, of course, the well pump that pumps the water into the house, but the pump cannot do the job alone.

In addition to a pump, most well systems rely on pipe, valves, pressure tanks, electricity, and other items to function properly. As a plumber, you must know how to install all aspects of these systems, with the exception of making the electrical connections.

Some pumps sit on top of their pressure tanks; others are suspended in the well water from a rope or pipe. Pumps may be located outside of the home, in pump houses. You might be working with a submersible pump or a jet pump. There could be one pipe in the well system or two. The electrical voltage may be 110 or 220 volts.

The pressure tank can be large or small, and it may or may not have a diaphragm in it. You will have to install a relief valve in the system, and a pressure gauge is standard equipment in a well setup. Much of the decision about what type of well system to install will depend on the type of water source you are pumping from.

There are three common types of wells: drilled, dug, and driven. The two most common are drilled and dug. The differences between a dug and a drilled well are their depth and the diameter of the well casing. Typically, a submersible pump is used in a drilled well and a jet pump is used with a dug well.

Already, you should be able to see that well systems are not simple. However, they are easy to understand and install. This chapter is going to explain well systems to you and show you how a system is installed.

Under normal conditions you will either be working with a dug well or a drilled well. We will begin our installation instructions with a dug well. Dug wells have large diameters and are often called shallow wells. The diameter of the well is large enough for a person to be lowered down the shaft, and the depth rarely exceeds 35 feet.

In the old days, dug wells were often lined with rocks, and plumbers would climb down the rock walls to work in the wells. Today, however, most dug wells are lined with concrete cylinders, and plumbers pull the piping and fittings out of the well to work on them.

The piping used in old wells was normally galvanized steel pipe. Modern wells normally contain polyethylene pipe: the black plastic pipe that comes in large coils.

The large diameter, usually around 3 feet, of shallow wells allows them to develop a reserve of water. Since the well is not deep, it must be wider to allow water to build up in the well casing. It is not unusual for shallow wells to run out of water during hot, dry summers; this is one reason many people prefer a drilled well.

Some shallow wells contain submersible pumps, but most of them deliver their water through a jet pump. Jet pumps are also called shallow-well pumps (Fig. 11.1). Shallow-well pumps may be fitted with one or two pipes. For wells with water depths of less than 25 feet, single-pipe pump systems are normally used.

These single-pipe pumps lift water with suction. The pump runs and creates a vacuum that pulls water up the well pipe. The maximum lift on this type of system is 30 feet, but it is recommended that they not be used at depths of more than 25 feet, since a perfect vacuum cannot be counted on.

Installing a Shallow-Well System

Installing a shallow-well system requires several types of fittings and equipment. For this lesson, let's start in the well, at the end of the pipe.

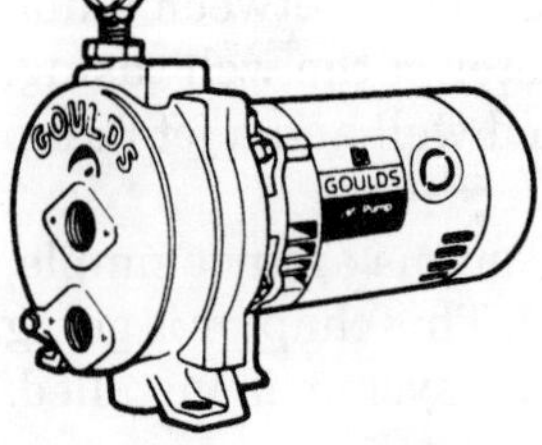

Figure 11.1 Jet pump. (*Goulds Pumps, Inc.*)

When you are ready to install your well system, you will need to have a trench dug from the well casing to the pump location. The pump may be installed inside habitable space, in a cellar, in a crawlspace, or in a well house.

Once the trench is open, you will need a hole in the side of the well casing. A rotary hammer-drill makes quick work of this job, but a star drill or cold chisel is often used to make the hole in the concrete casing. Once you have the hole and the trench, you are ready to begin your installation.

Okay, you have an open trench leading from the proposed pump location to the well casing. There is a hole in the well casing at a depth equal to the bottom of the trench, below the frostline. You have removed the large, heavy concrete cover from the well and are ready to install your piping.

Assuming that the well's depth will not require pumping water from deeper than 25 feet, you are going to install a suction system. If you need to check the depth of the well, you can do so by tying a nut or other weighted item to the end of a string and lowering it into the well. When it reaches bottom, retrieve the string and measure the depth.

The pipe you will be putting in the well and trench will have a diameter of 1¼ inches, and it should be rated for 160 pounds per square inch of pressure. You could get by with a lower rating, but 160 pounds per square inch is what most plumbers use. The pipe is polyethylene (PE), and you must be careful not to kink it.

Begin your installation by unrolling the pipe. Measure off the amount of pipe you will need and cut it with a hacksaw. As you work with the pipe, be careful not to bend it too acutely. If you do, the pipe will kink.

The fittings used with polyethylene pipe are called insert fittings (Fig. 11.2). These fittings have a shaft on one end that is inserted in the

Figure 11.2 Elbow insert fitting.

pipe. Ridges on the shaft allow the pipe to grip the fitting when it is inserted. The fittings are secured in the pipe with the use of stainless steel clamps. It is standard procedure for experienced plumbers to use two clamps at each fitting location.

The insert fittings may be plastic or metal. Either will work, but most plumbers prefer brass fittings for well installation. These fittings will not crack or break as easily as their nylon cousins.

Most shallow wells will contain a foot valve (Fig. 11.3). Foot valves are fittings attached to the end of the pipe in the well that filters, to some extent, the water being pulled into the pipe. These devices prevent sand and gravel from being sucked into the well pump. It is important not to place the foot valve too close to the bottom of the well. It should be suspended high enough from the bottom to avoid sucking up gravel, sand, and debris.

The foot valve is made to accept male threads. Start your installation by applying pipe-thread compound (pipe dope) to the male threads of a male-by-insert fitting (Fig. 11.4). This fitting will have male threads on one end and a ridged shaft on the other end.

Screw the male-by-insert fitting into the foot valve until it is tight. Now slide two stainless steel clamps over the end of your well pipe. Next, push the insert end of the male-by-insert fitting into the pipe. If the pipe is cold, you may have to warm it with your torch to get the fitting seated in it. Be careful though; if you hold your flame too close to the pipe, the polyethylene will melt or burn. Wave the flame over the pipe in sweeping motions to avoid burns and melting.

Once the insert fitting is inserted fully in the pipe, slide the clamps down over the pipe to the point where they encircle the shaft of the fitting. Tighten the clamps. Again, if the pipe is cold, you may have to heat it to get the clamps tight. Now you have one section of your piping ready for installation.

The next phase of your job is to roll pipe out in the trench, between the well and the pump. Stick one end of the pipe through the well casing and lay the pipe in the trench as you roll it out, toward the pump.

Return to the pipe with the foot valve on it and install an insert elbow

Figure 11.3 Foot valve. (*Goulds Pumps, Inc.*)

Figure 11.4 Male-by-insert adapter and female copper adapter.

fitting in the open end of the pipe. Clamp the fitting into place and prepare to join the other end of the elbow fitting with the pipe sticking through the well casing. Lower the foot valve into the well and make the connection with the pipe coming through the casing. This connection can be a little tricky.

Depending upon the depth of entrance for the pipe, you may not be able to reach over the side of the well casing to the pipe. In these cases, you have two options. You can lower a hook-type ladder into the well and work from it, or you may be able to fish the pipe up to within your reach and make the connection. If you opt to fish the pipe up for the connection, be careful not to kink it. Should you decide to climb down in the well, use a safety harness to assure you won't fall down the shaft.

Once the two pipes are connected with the elbow fitting, your plumbing work in the well is done. However, you will have to come back after securing the pump-end of the pipe and seal the hole in the well casing. This is normally done with a mortar mix, and the job is important. If you fail to seal the hole, nonpotable ground water can enter the well around the pipe.

Move along the trench with a shovel and lay dirt over sections of the pipe to hold it in place. Don't conceal the pipe; a plumbing inspector will want to look at it before it is covered completely. When you do fill in the trench, don't allow sharp rocks to cut the pipe. Fill the trench with clean fill dirt. Now, relocate your tools and materials to the pump location; that is where the rest of the work will take place.

The jet pump could be located in many places, but for this example it is going to be installed in the basement of a home. The supply pipe from the well must be brought into the basement. This normally requires making a hole in the basement wall. If you are working with a good general contractor, the foundation crew will have left a sleeved access hole for you. If not, you will have to make your own.

In block and brick foundations, you can make your hole with a star drill, a rotary hammer, or a cold chisel. For thick concrete walls, you may need a jackhammer. In any event, make the hole, and make it large enough to accommodate a pipe sleeve. The sleeve can be a piece of polyvinyl chloride (PVC) drain pipe, but it must be at least two pipe sizes larger than the water service pipe. Install the sleeve and parge the hole to seal around the sleeve.

Run your water service through the hole. Once inside the house, you will probably convert the PE pipe to an acceptable interior water distribution pipe, like copper tubing or polybutylene (PB). In this example, we are going to work with copper tubing.

To convert the PE pipe to copper, you will use a male-by-insert adapter. Start the conversion by soldering a female adapter onto a length of copper tubing. Install the insert fitting into the PE pipe and clamp it. Apply pipe dope to the male threads, and screw the female copper adapter and length of tubing onto them. Your conversion is made and the remainder of your piping will be copper.

Once your water service is secured and converted, install a gate valve near where the pipe enters the home. Next install a check valve in the line to prevent water from running out of the house and back down the well. If you installed a foot valve in the well, you may omit the in-line check valve. The foot valve will act as a check valve. Extend the main line to the pump.

Jet pumps can be mounted on boards that sit on the floor, on pressure tanks, or just about anywhere else you would like to mount them. The pump will be tapped to accept male threads. Dope up a male adapter and screw it into the inlet opening on the face of the pump. Your main incoming water line will attach to the male adapter.

The next step will be to set the pressure or holding tank in place. Small versions of these tanks sometimes mount in-line, on the outgoing water pipe of the pump. Normally, however, they sit on the floor, near the pump. Some models have brackets that allow the pump to be mounted on top of them (Fig. 11.5). Most modern tanks are captive air tanks. The diaphragms found in most new pressure tanks all but eliminate a condition known as waterlogging.

Old holding tanks, without diaphragms, were charged with air to allow the water to build pressure. Over time, this air would dissipate, causing the tank to fill with water and little air. This results in a lack of water pressure and is known as waterlogging.

Once the tank is in place, install a tank tee. This is a special fitting that screws into the pressure tank and allows a pressure gauge, control switch, drain valve, and relief valve to be mounted on it.

Next, dope up and screw a male adapter into the outlet hole of the pump. Install a short piece of pipe and a union fitting. Then extend pipe from the opposite side of the union to the tank tee. This gets you to the tank.

Install the accessories on the tank tee: the pressure gauge, relief

Figure 11.5 Tank-mounted jet pump. (*Goulds Pumps, Inc.*)

valve, drain valve, and pressure switch. Come off the unused end of the tank tee with the pipe that will distribute cold water to the home's plumbing system. Install a gate valve near the pressure tank and extend your water main to the fixtures in the home.

If you choose not to use a tank tee, connect to the pressure tank in the following manner (Fig. 11.6). Come out of the outlet and install a tee fitting. Use reducing fittings on the top of the tee to size it for accepting a pressure gauge. From one side branch of the tee, run a short piece of pipe and install a union. Continue from the union to the pressure tank. A pressure relief valve must be installed in a tee fitting, near the pressure tank.

From the other side branch of the tee with the pressure gauge on it, install a short piece of pipe and a union. Come out of the other side of the union with another short piece of pipe and install a gate valve. From there you can go on to supply the fixtures in the home.

This type of installation will have the pressure switch mounted on the pump. When the electrical connections are made, you are ready to prime your pump.

Remove the prime plug from the top of the pump and pour water into it. You must pour in enough water to fill the pipe from the pump to the foot valve. When water is standing in the priming hole, apply pipe dope to the threads of the prime plug and replace it. Turn the pump on, and you should get water pressure. If you don't, repeat the priming process.

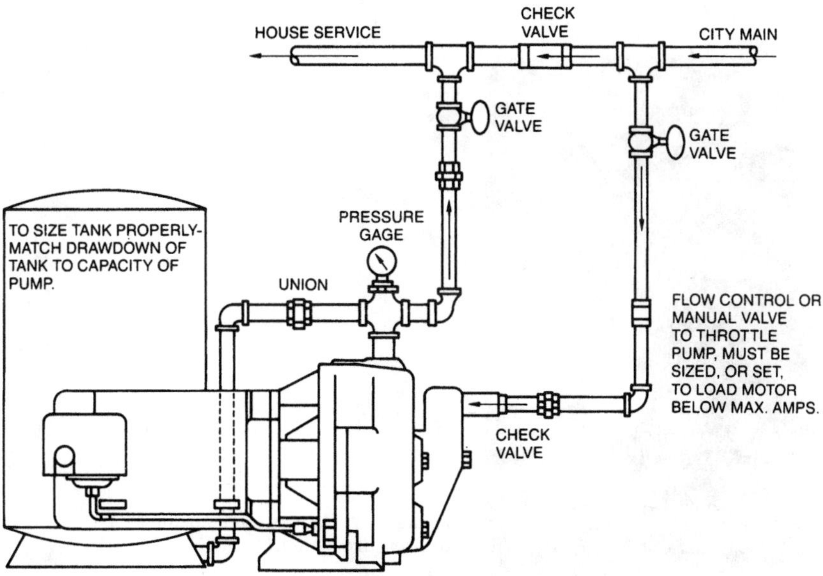

Figure 11.6 Typical jet pump installation. (*Goulds Pumps, Inc.*)

When the pump produces water, the pressure tank will begin to fill. Once the tank is full, the pressure gauge should show a pressure of between 40 and 60 pounds per square inch. If the pressure is lower or higher, you should adjust the pressure switch.

The switch will be preset with cut-in and cut-out pressures. Normally, the cut-in pressure will be around 20 pounds per square inch. This setting tells the pump when to cut on and pump more water. When the pressure in the tank is low enough to activate the cut-in switch, the pump cuts on automatically.

The cut-out setting is the pressure setting that tells the pump to cut off. For example, if you want 50 pounds per square inch in the tank, the setting would be set at 50 pounds per square inch. When the tank reaches this pressure, the pump will cut off.

Both the cut-in and cut-out settings are adjustable. By removing the cover on the pressure switch, you will expose springs and nuts (Fig. 11.7). You will also be exposing electrical wires that could shock you. By loosening and tightening the nuts that compress the springs, you can adjust the settings for cut-in and cut-out pressures.

Deep-Well Jet Pumps

In many respects, deep-well jet pumps are like shallow-well pumps. There are, however, differences in the way they are installed (Fig. 11.8). Both types of pumps look about the same. In fact, deep-well

Figure 11.7 Adjustment springs in pressure switch.

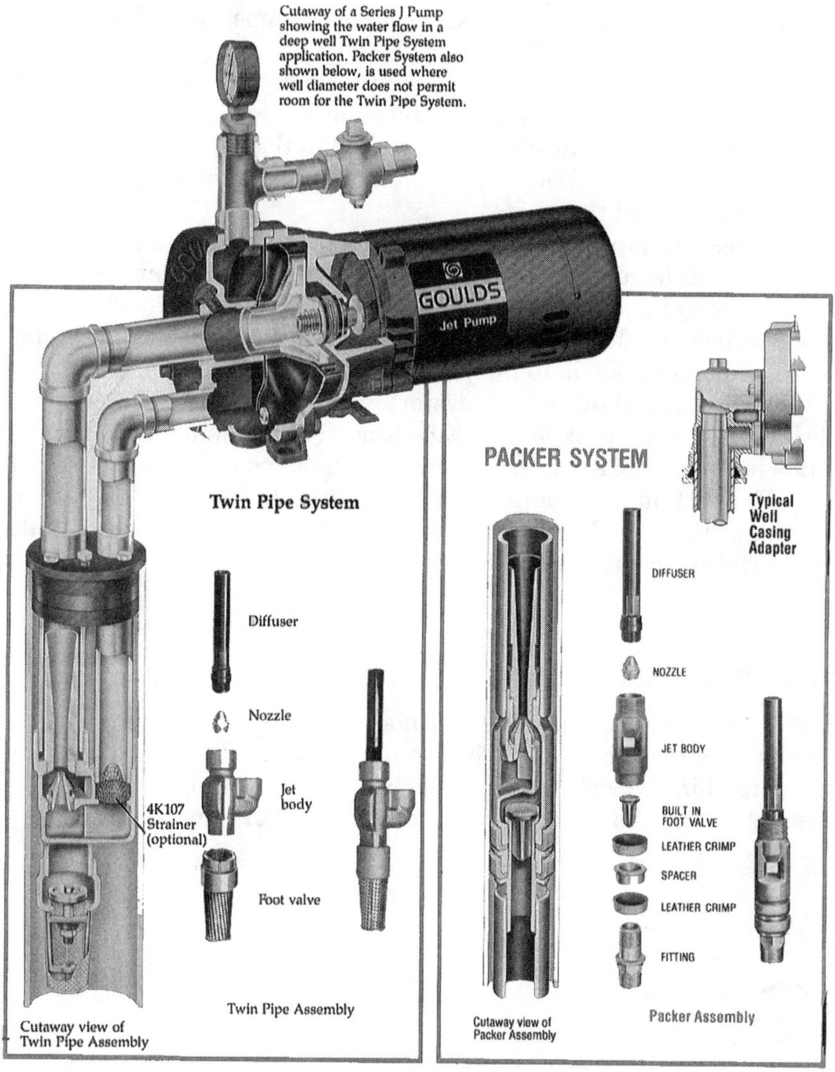

Figure 11.8 Deep-well jet pump installation. (*Goulds Pumps, Inc.*)

pumps are often used as shallow-well pumps once a shallow-well adapter is added. The most noticeable difference between the two types of systems is the number of pipes going into the well. A deep-well system uses two pipes instead of just one.

These pumps work by pushing water up out of the well rather than sucking it up. This allows the pump to work with wells having depths of more than 30 feet. However, most modern pump installations for wells deeper than 25 feet use submersible pumps; we will talk more about them in a little while.

The piping arrangement from the outlet hole of a two-pipe pump is the same as described for a shallow-well pump except for the addition of a pressure control valve that is installed in the incoming pipe before it reaches the pump. The major differences lie in the installation between the well and the pump.

The pressure pipe will normally have a diameter of 1 inch. The suction pipe will have a diameter of 1¼ inches. In this system, the pump runs, pushing pressure down the pressure pipe to a jet body that hangs in the well water. Then the suction pipe, which is connected to the same jet body, pulls water up to the pump.

When you install this type of system, you will install a diffuser, a nozzle, and a jet body that were not mentioned in the shallow-well installation instructions. All of these items are installed on the pressure pipe side. The suction pipe connects to a fitting on the jet body. All of these parts are submerged in the well water. The remainder of the installation is the same as for a shallow-well pump.

Submersible Pumps

Submersible pumps (Fig. 11.9) are normally used with wells that have depths of more than 25 feet. These pumps hang at the bottom of a pipe, submerged in the well water. Submersible pumps are more expensive than jet pumps, but in many ways they are better pumps.

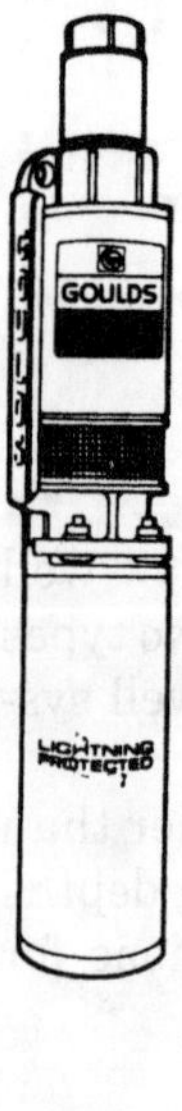

Figure 11.9 Submersible pump. (*Goulds Pumps, Inc.*)

When a submersible pump's performance is compared to a jet pump of equal horsepower, the results are astonishing. A ½-horsepower submersible pump is capable of providing about 300 more gallons of water than a ½-horsepower jet pump. Submersible pumps don't require foot valves or suction lines. They hang in the well water and push water up to the pressure tank.

Many of the wells that use submersible pumps have very small diameters: often only 6 inches across (Fig. 11.10). The casings for drilled wells are made of steel. When you want to install a submersible pump, you must cut a hole in the steel casing below the frostline. This can be done with a cutting torch or a quality hole saw.

Once you have your trench open and your hole cut in the casing, you are ready to begin your installation (Fig. 11.11). The first order of business is a pitless adapter. This is a device that mounts to the side of the well casing and holds the pump suspended in the water. The pitless adapter is in two parts. One part attaches to the outside of the well casing and the other part to the inside. When the device is tight, it seals the hole in the well casing. The pipe in the trench attaches to the pitless adapter and is run to the pressure tank.

The rest of the piping and pump work will be concealed in the well. All threaded joints should be prepared with pipe dope on the male threads. You will use the same PE pipe used with other well installations. Before prefabbing your pump setup, establish the depth of the

Figure 11.10 Drilled well.

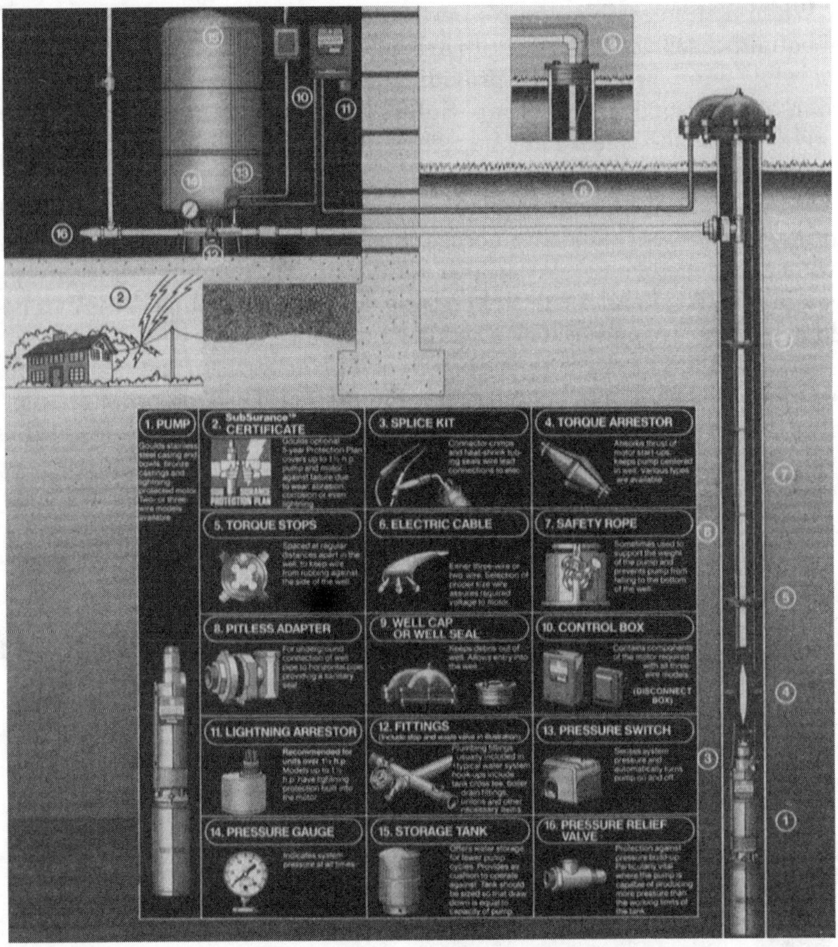

Figure 11.11 Submersible pump installation. (*Goulds Pumps, Inc.*)

well. Before you can size the proper pump for the well, you must know the depth and flow rate of the well. The company that drilled the well should be able to provide you with this information.

Submersible pumps are rated for flow rates and depths. If the pump you install is too powerful, it can pump the well dry. If it is not powerful enough, it will not get water to the pressure tank. There are sizing charts available from pump manufacturers to help you in sizing your jobs.

Start fabricating your setup with the pump. When you cut your piece of well pipe, plan for the pump to hang at least 10 feet above the bottom of the well and at least 10 feet below the lowest anticipated water depth.

Install a male insert adapter into the top of the pump. A torque arrester and two pipe clamps should then be slid over the well pipe. Slide the end of the pipe over the insert fitting and secure the pipe with the clamps. The torque arrester can be adjusted on the pipe to make it balloon out. It should be expanded to a size just smaller than the well casing and should be in close proximity to the pump. This device protects the pipe and wires from abrasion when the pipe vibrates during use, and it keeps the pump centered in the well.

Now slide torque stops down the well pipe from the end opposite the pump. These items also help protect against abrasion, and they provide a means for securing the wires feeding electricity to the pump. The torque stops should be placed at regular intervals along the pipe.

When you connect the pump wires to the electrical service wires, you must use waterproof splice kits. Many plumbers wire their own pumps and only rely on electricians to make the connection at the service panel.

When the wiring connections are made at the pump, the wiring is wrapped around the pipe and taped in place, or it is laced in the torque stops and taped to the pipe. The wires are run up to the top of the well casing and connected to the underground wire running from the service panel to the well casing.

The next step is to screw an insert-by-male adapter into the pitless adapter. Once this is done, you can install the insert fitting in the top end of the pipe. Not all plumbers do so, but you should tie a rope onto the water pump before lowering it into the well. If the pipe connections should come loose, the pump can be retrieved with the rope. Otherwise, the pump will be lost in the well.

When you look at a pitless adapter, you will see it is basically a tee fitting. The center of the tee goes to the water service pipe. The bottom of the tee is where the well pipe connects, and the top of the tee is used to mount and dismount the adapter.

Most plumbers have a length of steel pipe that they use to raise and lower pumps. One end of the steel pipe is screwed into the top of the pitless adapter. The other end is often fitted with a tee and short pieces of pipe to form a T-handle. When this tool is screwed into the top of the pitless adapter, it can be used to guide the adapter into place. Once the adapter is slipped into place and pushed down tight, the steel pipe can be unscrewed and used for other jobs.

Raising and lowering submersible pumps is a job best done by at least two people. Special care is required to avoid damaging the electrical wiring on the well casing. If the insulation on the wires is damaged, many problems can arise.

Once all the well connections are complete and you are inside the foundation, the plumbing connections are somewhat similar to the ones made with jet pumps.

Figure 11.12 Diaphragm-type pressure tank. (*Goulds Pumps, Inc.*)

A pressure tank (Fig. 11.12) is set in place and a tank tee is installed. The tank tee will support the pressure switch, the pressure gauge, and a drain valve. A pressure relief valve must also be installed, and, of course, gate valves will be installed on the incoming pipe and the pipe feeding the fixtures in the home.

Installing a well system is not a complicated process. Once you have done it a few times, it becomes easy. As an apprentice, you will be working alongside a plumber, learning first hand how the job is done. Now we are ready to move on to Chapter 12 and check into septic systems.

Chapter

12

Septic Systems

Do you ever think much about septic systems? Don't feel bad, neither do most other people. But the way septic systems work is quite interesting, and you might enjoy learning how bacteria is working day and night in a septic tank. You would probably be surprised to know how simple it is for a septic system to modify raw sewage and turn much of it back into drinking water. That's right, septic systems do take sewage and convert much of it to groundwater that later becomes drinking water.

Did you know there are different types of septic systems? Were you aware that what people put down their drains can damage a leach field? For that matter, do you know what a leach field is and how it works? Well there are different types of systems, and what people put down their drains can have an adverse effect on the environment.

Should homes with private septic systems be equipped with garbage disposers? This is a debatable subject. Some jurisdictions prohibit the installation of a garbage disposer when a private waste disposal system is in use and others don't.

Do plumbers really need to know anything about septic fields? Well, plumbers don't normally install septic systems, but they are often the first people called when a septic system fails. There are many aspects of septic systems that plumbers should understand. And that is what this chapter is all about. You are about to get a lesson in septic systems and how they affect plumbers.

What Is a Septic System?

A septic system is a group of components that work together in the treatment of sewage. A typical residential septic system will include a septic tank, a distribution box, solid and slotted pipe, and a leach field. The layout of these components begins with the leach field.

How Are Septic Systems Installed?

Before a septic system can be installed, the ground conditions must be inspected to determine if the soil meets local health requirements for sewage treatment. Soil samples are taken with augers by county health officials or private engineers and evaluated. The soil is checked for its percolation rating.

Once the best location on the building lot has been determined, the layout of the septic system is designed. This design might be done by local authorities or by private engineers. Once the septic system is designed and approved, a permit is solicited. When the permit to install the system is issued, the work may begin.

Plumbers extend a home's building drain beyond the foundation, by about 5 feet, where it becomes the sewer. The company installing the septic system installs the septic tank. Average residential jobs have concrete septic tanks with a capacity of 1000 gallons. The tank is buried in the ground.

Most septic tanks have two access holes in their tops. The holes are covered by removable lids, usually made of concrete. The holes allow access to pump the tank out and to work on the inner portions of the tank.

Once the tank is installed, plumbers extend the sewer from the house to the tank and make the connection at the inlet of the septic tank.

The septic installers do the rest. They bury a small box, called a distribution box, a short distance beyond the septic tank. A solid piece of 4-inch pipe is installed between the septic tank and the distribution box.

From the distribution box, numerous pipes branch off and go into the leach field. These pipes are slotted and are usually bedded in gravel. The number of pipes going into the leach field is determined by the septic design. The length of the pipes is also specified in the septic plan. Once all the work has been inspected and approved by the proper authorities, the system is covered with dirt.

In areas where soil quality is poor, chambered systems are used. These systems differ from standard systems in the construction of the leach field. Instead of 4-inch slotted pipe, special chambers are used. The chambers may be made of concrete or plastic. The increased size of the chambers compensates for the slower absorption rate of the poor soil.

How Does a Septic System Work?

A standard septic system works on a very simple principal. Sewage from a home enters the septic tank through the sewer. Where the sewer is connected to the septic tank, there is a baffle on the inside of the tank. This baffle is usually a sanitary tee. The sewer enters the center of the tee and drops down through the bottom of it. The top hub of the tee is left open.

The bottom of the tee is normally fitted with a short piece of pipe. The pipe drops out of the tee and extends into the tank liquids. This pipe should never extend lower than the outlet pipe at the other end of the septic tank. The inlet drop is usually no more than 12 inches long.

The outlet pipe for the tank also has a baffle, normally an elbow fitting. The drop from this baffle is frequently about 16 inches in length.

When sewage enters a septic tank, the solids sink to the bottom of the tank and the liquids float within the confines of the container. As the tank collects waste, several processes begin to take place.

Solid waste that sinks to the bottom becomes what is known as sludge. Liquid, or effluent as it is called, is suspended between the lower layer of sludge and an upper layer of scum. The scum layer consists of solids and gases floating on the effluent. All three of these layers are needed for the waste disposal system to function properly.

Anaerobic bacteria works inside the septic tank to break down the solids. This type of bacteria is capable of working in confinements void of oxygen. As the solids break up, they form the sludge layer.

As the effluent level rises in the tank, it eventually flows out of the tank, through the outlet pipe. The effluent drains down a solid pipe to the distribution box. After entering the distribution box the liquid is routed into different slotted pipes that run through the leach field.

As the effluent mixes with air, aerobic bacteria begin to work on the waste. This bacteria attacks the effluent and eventually renders it harmless. Aerobic bacteria needs oxygen to do its job. Drain fields should be constructed of porous soil or crushed stone to ensure the proper breakdown of the effluent.

As the effluent works its way through the drain field, it becomes odorless and harmless. By the time the effluent passes through the earth and becomes groundwater, it should be safe to drink.

Septic Tank Maintenance

Septic tank maintenance is not a time-consuming process. Most septic systems require no attention for years at a time. However, when the scum and sludge layers combined have a depth of 18 inches, the tank should be cleaned out.

Trucks equipped with suction hoses are normally used to clean septic tanks. The contents removed from septic tanks can be infested with germs. The disease risk of exposure to sludge requires that the sludge be handled carefully and properly.

What Happens if the Drain Field Doesn't Work?

When a septic field fails to do its job, a health hazard exists. This situation demands immediate attention. The main reason for a field to fail in its operation is clogging.

If the pipes in a drain field become clogged, they must be excavated and cleaned or replaced. If the field itself clogs, the leach bed must be cleaned or removed and replaced. Neither of these propositions is cheap.

How Can Clogs Be Avoided?

Clogs can be avoided by careful attention to what types of waste enter the septic system. Grease, for example, can cause a septic system to become clogged. Bacteria does not do a good job of breaking down grease. Therefore, the grease can enter the slotted drains and leach field with enough bulk to clog them.

Paper, other than toilet paper, can also clog up a septic system. If the paper is not broken down before entering the drain field, it can plug up the works.

Do Garbage Disposers Hurt a Septic System?

Many people, including numerous code enforcement offices, believe garbage disposers should not be used in conjunction with septic systems. Other people disagree and believe that disposers have no adverse effect on a septic system.

It is possible for the waste of a disposer to make it into the distribution pipes and drain field. If this happens, the risk for clogging is elevated. Another argument against disposers is the increased load of solids they put on a septic tank. Obviously, the amount of solid waste will depend on the frequency with which the disposer is used.

My opinion is that disposers increase the risk of septic system failure and should not be used with such systems. However, I know of many houses that are using disposers with septic systems and are not experiencing any problems. If you check with your local plumbing inspector, this question may become a moot point. Many local plumbing codes prevent the use of disposers with septic systems.

Piping Considerations

There are some additional piping considerations for plumbers to observe. Septic tanks are designed to handle routine sewage. They are not meant to modify chemical discharges and high volumes of water. If, as a plumber, you pipe the discharge from a sump pump into the sanitary plumbing system, which you are not supposed to do, the increased volume of water in the tank could disrupt its normal operation.

Chemical drain openers used in high quantities can also destroy the natural order of a septic tank. Chemicals from photography labs are another risk plumbers should be aware of when piping drainage to a septic system.

Gas Concentrations

Gas concentrations in a septic tank can cause problems for plumbers. The gases collected in a septic tank have the potential to explode. If you remove the top of a septic tank with a flame close by, you might be blown up. Also, breathing the gases for an extended period of time can cause health problems.

Sewage Pumps

There are times when sewage pumps must be used to get sewage to a septic system. The pumps are normally installed in a buried box outside of the building being served. The box is often made of concrete. In these cases, the home's sewer pipe goes to the pumping station. From the pumping station, a solid pipe transports the waste to the septic tank.

Sewage pumps must be equipped with alarm systems. The alarms warn the property owner if the pump is not operating and the pump station is filling with sewage. Without the alarm, the sewage could build to a point where it would flow back into the building.

Sewage pumps have floats that are lifted as the level of contents in the pump station build. When the float is raised to a certain point, the pump cuts on, emptying the contents of the pump station. The discharge pipe from the pump must be equipped with a check valve. Otherwise, gravity would force waste down the pipe, back into the pump station, when the pump cut off. This would result in the pump having to constantly cut on and off, wearing out the pump.

Plumbers' Pointers

Let me give you some plumbers' pointers on septic tanks. If a property owner is complaining of foul odors outside a home, check the septic system. In all likelihood, the septic field has either quit working or the system has been flooded with too much water.

If you arrive at a home that has sewage backed up in the plumbing fixtures, check the septic tank. This could save you a lot of time and trouble. If the liquid level in the tank is above the inlet pipe, have the property owner get the tank pumped out.

If you must work in close proximity to the contents of a septic tank, wear protective clothing and gloves. The diseases you might get from coming into direct contact with sewage are not nice.

This has been a short chapter, but you have been told all you should need to know about septic tanks. Since there is no sense wasting time and space on useless words, let's move on to the next chapter and take a look at water conditioning systems.

Chapter

13

Water Conditioning Systems

Water conditioning systems are fast becoming standard equipment in many homes. As water quality continues to get worse and consumers become better educated, the demand for water conditioning systems is increasing. The demand is so great that many companies provide services specializing in the treatment of water. It is not unusual to look through a phone directory and find several companies that do nothing but water conditioning.

Plumbers frequently install, maintain, and repair water conditioning equipment, but the field has become flooded with specialists. As a working plumber, you may never have a need to work on a water softener or acid neutralizer, but you might. In fact, you might want to specialize in the conditioning and treatment of water.

This chapter is going to discuss the types of water conditioning equipment available and explain, in basis terms, how the equipment is installed. If you will be working in rural areas, where private water systems are common, you are very likely to run into water conditioning equipment. Even if your work is confined to municipal water supplies, you will see that there are instances when the water might be better after additional treatment. Now, let's see what water conditioning is all about.

Why Does Water Need to Be Conditioned?

There are a multitude of reasons for treating and conditioning water. Since there are so many reasons, let's look at them one at a time.

Sulfur

When water contains a concentration of sulfur, it tends to smell like rotten eggs. Water with sulfur in it may take on a blackish tint. This impure water can damage plumbing fixtures and equipment. The taste and odor of water with sulfur in it is usually quite unpleasant.

Hard water

Hard water is a common problem in households; it contains calcium and magnesium. These minerals can cause build-ups on the elements of electric water heaters that render them nearly useless. If you ever remove a water heater element and see a build-up of white, crusty material, you are looking at the result of hard water.

Another problem associated with hard water is the use of soap and detergents. The nature of hard water makes it all but impossible to enjoy a good lather from soap or shampoo. The fact that hard water does not work well with soap products results in clothes that won't come clean, dishes that develop filmy residues, and hair that won't respond well to shampooing.

Deposits left on the inside of plumbing pipes by hard water can restrict water flow and result in reduced water pressure. It is possible for hard water to cause plumbing pipes, fittings, fixtures, and equipment to deteriorate prematurely.

Acidity

Acidity is a common problem for people who use well water. Water with a high acid content can be damaging to plumbing pipes, fixtures, and equipment, as well as human health. Green stains in plumbing fixtures or unexplained pin-hole leaks in copper tubing are indicators of acidic water.

Iron

Iron content is frequently encountered with well water. Have you ever looked into a toilet tank and seen a black layer of particles on the bottom of the tank? These black particles are proof of a high iron content in the water. Iron in potable water can cause an unpleasant taste and stained laundry. A slime buildup in your plumbing fixtures is another indicator of iron being present in the water system.

Diseases

Water that is not properly treated can result in the contraction of many diseases, some of which are very serious. A list of the diseases commonly associated with impure water could include diarrhea, cholera, typhoid fever, paratyphoid, hepatitis, and schistosomiasis. Some of the diseases can be as dangerous as they are hard to pronounce.

The reasons listed are the most common motivators for installing water conditioning equipment. The equipment used to treat water for these undesirable contents can vary a great deal. Let's now turn our attention to the types of conditioning equipment used to combat bad water.

Types of Water Conditioners

There are many types of water conditioners. We will look at each of the most common types in this section. Again, we will examine them one by one.

Water softeners

Water softeners are probably the best known type of water conditioning equipment. These devices are used to treat hard water. They are occasionally used to treat minor cases iron content.

Unlike a filter through which the water passes and is cleaned, water softeners must go through different cycles in treating hard water. The first stage of treatment is known as a backwash cycle. When this phase is complete, the unit enters a regeneration phase. A rinse cycle follows the regeneration phase. Once all the cycles are complete, the equipment returns to its ready mode.

During the softening procedure, water is passed through a mineral bed. This bed absorbs the calcium and magnesium contained in the water through an ion exchange. Sodium is released during this phase of treatment. Iron is also removed during this part of the operation.

Synthetic minerals are often used in water softeners. These materials are capable of absorbing a higher quantity of calcium carbonate than natural materials can. The mineral bed does not dissolve and can be used over and over again. However, when the mineral bed has absorbed its capacity of particles, it must be regenerated.

The regeneration process occurs with the use of ordinary salt. The salt is dissolved in the water. The brine created by the salt forces the calcium and magnesium to leave the mineral bed.

There are various types of water softeners. Some units depend on the routine replacement of tanks to remain in operation. Others rely on the addition of salt as a maintenance procedure. Because of the number of types of water softeners available, it is necessary to read and follow the manufacturer's instructions in the installation and maintenance of the equipment.

Carbon filters

Carbon filter are quite common, and they work on simple principals. These filters use carbon particles to filter disagreeable tastes, odors, and fine particles from water. You can find carbon filters in many different applications. They may hang inside a water bottle to filter water being poured into the bottle. Carbon filters are often mounted on the aerator threads of faucets. And you can find carbon filters that mount in-line on water pipes.

Acid neutralizers

Acid neutralizers neutralize acid. There are two basic types of acid neutralizers. The first type works with calcium carbonate, and the other type works with soda ash.

Sand filters

Sand filters trap floating particles in water. They control silt and mud, resulting in clearer water. Sand filters are not common household filters.

Chlorinators

If a water system is plagued by bacteria and algae, a chlorinator will probably be used to solve the problem. Residential systems frequently work with any type of average laundry bleach.

Iodinators

Iodinators work to control bacteria, fungi, and viruses. If an iodinator fails to do its job, big trouble can result.

Installing Water Conditioning Equipment

Installing water conditioning equipment is not much different from any other type of plumbing. However, it is vital that you follow the manufacturer's recommendations for a proper installation. Some systems use a complicated maze of piping that is best described by the manufacturer. Other types of equipment, such as charcoal filters, might simply screw onto the aerator threads of a faucet. Here we are going to take a look at some basic installation methods, but you should always refer and adhere to the manufacturer's recommendations when installing water treatment systems and devices.

Screw-on filters

Screw-on filters don't need much explanation. You simply remove the aerator from a faucet and screw the filter on in its place.

In-line filters

In-line filters are common, and the installation of these units is simple. First, cut the water off to the pipe where the filter will be installed; normally this is the main water distribution pipe. With the water off, cut out a section of water pipe. Next, install valves on each end of the remaining pipe. The filter should be isolated with a valve on each side of it. Since these units require occasional filter changes, it is helpful to have valves close by.

Most in-line filters have clear housings that hold their filters. The housing should have an inlet and an outlet hole. Both holes should be threaded to accept male adapters. All you have to do is install the filter between the two valves, with the inlet and outlet holes in the proper locations. That's really all there is to it.

Complex water conditioning equipment

Complex water conditioning equipment (Fig. 13.1) is a bit more complicated to install. When you get into sophisticated equipment, you will have to rely on the manufacturer's recommendations. Most of the materials used for the installation will be standard plumbing materials, but the locations of valves and various devices can be critical to a system's operation.

Some systems will be very simple in their connection requirements,

The Iron Curtain™ Filtration System

Hellenbrand's Iron Curtain™ Filtration System is the single system that guards against a multitude of water problems. Imagine. No more iron taste. No more sulfur smell. No more corroding pipes. Longer life for your appliances, water heater and softener. Fresh, clean smelling, great tasting water can be yours again with the Iron Curtain™ filtration system.

Our Revolutionary Design Means Efficient Operation

The Iron Curtain™ is unlike old fashioned iron filters. Its unique operation reduces costs and maintenance while delivering years of trouble-free service.

The Iron Curtain™

- Requires no chemicals or salt for operation or regeneration
- Requires no air injectors, venturis or micronizers
- Removes dissolved and precipated iron in high concentrations
- Automatically increases pH of acid water on type "A" Systems

Operation is simple. Mineral rich water is forced through an aeration tank which converts dissolved minerals such as iron, manganese and sulfur into physical particles. These tiny mineral particles then circulate in the system for enough time to grow to a size where they are trapped in the filter. This way the highest level of dissolved minerals can be easily removed from your water.

No Obligation Water Analysis

There is a Hellenbrand Iron Curtain™ Filtration System for virtually any residential or commercial need. Your Hellenbrand dealer will be happy to test your water to determine the best system for you. Always have your water tested before you purchase.

Water Analysis

Iron ________________(p.p.m.) Hydrogen Sulfide ________(p.p.m.)
Manganese ________________(p.p.m.) pH ________________(number)
Tannins ________________(p.p.m.) TDS ________________(p.p.m.)
Iron Bacteria_______Yes_______No

TDS Limit 750 p.p.m. Tannin Concentration Limit .5 p.p.m. If either of these limits are exceeded the IRON CURTAIN is not recommended for this application.

IRON CURTAIN™ SYSTEM Specifications

Model	Filter Aeration Tank Size	Media Cu. Ft.	Inlet Outlet	Max. Service (1.) Flow GPM	Backwash Rate GPM	Floor Space (W x H x D)
IC-10 (1)	10" x 54"	1.5	3/4"	5.0	5.0	26" x 70" x 16"
IC-10A (2)	10" x 54"	1.5	3/4"	5.0	5.0	26" x 70" x 16"
IC-12	12" x 52"	2.0	1"	7.0	8.0	30" x 68" x 18"
IC-12A	12" x 52"	2.0	1"	7.0	8.0	30" x 68" x 18"

Consult factory for commercial applications.

1. Water temperature above 60°F will require higher backwash rates–Contact factory.
2. Validated by the Water Quality Association S-200 Standard for: Filtered water with less than .2 p.p.m. (Fe) Iron, .05 p.p.m. (Mn) Manganese, and .2 ppm (H2S) Hydrogen Sulfide when challenged with 10 p.p.m. (Fe) Iron, 2 p.p.m. (Mn) Manganese, and 10 p.p.m. (H2S) Hydrogen Sulfide respectively.
3. Validated by the Water Quality Association S-200 Standard for: Filtered water with greater than 6.8 pH when challenged with water of 6.0-6.2 pH, in addition to results in #2 above.
4. Actual operating conditions in Owners Manual may limit specific contaminants. See operating conditions in Owners Manual for details.

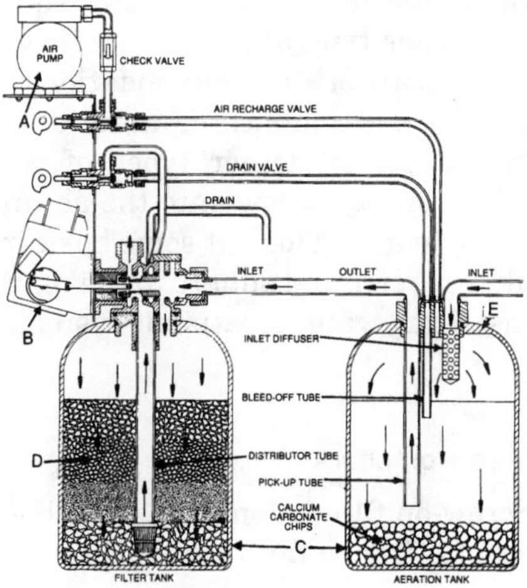

Parts Descriptions

A. 110 volt aeration pump automatically injects air into the aeration tank during the regeneration process without the help of injectors, venturis or micronizers.

B. Iron Curtain™ Control Center automatically controls the regeneration steps of backwash, aeration bleed-off, aeration recharge and final filter rinse before going into service. All cycles are adjustable with regeneration frequency controlled by a 12 Day Timer.

C. PolyGlass mineral tanks will not rust or corrode. National Sanitation Foundation listed.

D. Iron Curtain™ filter uses multiple layers of filtering media to provide longer filter runs, higher service flow rates and higher water clarity.

E. Iron Curtain™ Aeration System sprays raw water through a 18" head of air and then provides several minutes of additional contact time to allow the iron, manganese and/or hydrogen sulfide to fully oxidize. Special calcium carbonate chips are used to raise the pH and speed up the oxidation process.

Figure 13.1 Multipurpose water conditioner. (*Iron Curtain Filter System Parts Description. Source: Hellenbrand Water Conditioners, Inc. Waunakee, WI*)

and others will have you tracing diagrams for just the right fittings and valves. Pay close attention to what the manufacturer recommends, and you shouldn't have any trouble. Now, let's move on to Chapter 14 for a look at water services and sewers.

Chapter

14

Water Services and Sewers

All plumbing systems must have a water service and a sewer. The water service brings water to the system, and the sewer takes it away. It doesn't matter if a building is in a remote rural area or a major city; it is going to have a water service and a sewer if it has a modern plumbing system. While all plumbing systems depend on a water service and a sewer, the materials and installation procedures for these plumbing components can vary significantly.

This chapter is going to deal with all aspects of installing water services and sewers. You will learn about the materials used for these sections of a plumbing system, and you will learn what is required during their installation. The basic principles used for these installations are simple, but the code requirements complicate the issue. This chapter will remove the confusion sometimes associated with water services and sewers.

What Is a Water Service?

A water service is a pipe that brings water from a water source to a building. The water source can be a well or a municipal water main. The pipe that runs from a water source to the inside of a building's foundation is a water service. The water pipes within a building are called water distribution pipes. The main water pipe delivering water from outside a building to the water distribution pipes is the water service.

What Is a Building Sewer?

A building sewer is the pipe that begins outside a building's foundation and extends to a private waste disposal system or a municipal sewer system. The main drainage pipe found inside a building is called a building drain. It does not become a building sewer until it leaves the building.

What Materials Are Approved for Use as Water Services?

There are numerous materials approved for use as water services. You will have to check your local plumbing code to determine which materials are approved in your area, but I can give you a good idea of what is likely to be approved.

Copper

Copper pipe is an approved water service material. The types of copper that are usually approved include K, L, and M. Copper can also be used as a water distribution pipe.

Polyethylene (PE)

PE pipe is typically approved for water service applications. However, in many parts of the country, this type of pipe may not be used as a water distribution pipe. This means that the water service must be converted to an approved water distribution pipe material within the first 5 feet of its entry into a building.

Polybutylene (PB)

PB can be used for water services and as a water distribution pipe.

Polyvinyl chloride (PVC) and acrylonitrile butadiene styrene (ABS)

PVC and ABS pipe designed for use with potable water can be used as water services but may not be allowed for use as water distribution pipes.

Other approved materials

Other approved materials include galvanized steel, asbestos cement, brass, and cast-iron pipe. Before using any of the materials mentioned in this book, you should check your local regulations. Local code authorities can amend plumbing codes to suit their requirements, and the information in this book pertaining to code compliance may not be accurate in some sections of the country. The materials used most often for individual water services are polyethylene, copper, and polybutylene.

What Materials Are Approved for Use as Sewers?

Just as with water services, there are many materials approved for use as sewers. Those normally accepted for use are ABS, asbestos cement, bituminized fiber, cast-iron, concrete, copper, PVC, and vitrified clay. Again, regional differences may allow the use of pipe materials not listed or exclude the use of materials listed; check your local code before making any installations. The materials used most often for individual building sewers are PVC, ABS, and cast-iron pipe.

Now you have a general understanding of what materials are allowed for use as water services and sewers. The next step toward becoming a plumber is learning how to install these materials. That is precisely what we are about to get into.

Installing a Water Service

Installing a water service is not a particularly difficult job. In theory it is quite simple. You hook one end of the pipe to a water distribution pipe in a building and run the other end to a water source. While that is what happens, it doesn't happen that easily.

When we studied well systems, back in Chapter 11, you saw one way to install a water service. The pipe running from the well to the pump or pressure tank was the water service. If you were installing a water service in a city environment, you would be connecting to a water meter yoke.

Municipal hookups for water services start near the edge of the property line and run into the house, usually less than 5 feet inside the foundation. Once inside the foundation, the water service pipe is converted to a material approved for water distribution, if need be. The pipe must be sleeved as it passes under or through the foundation. The sleeve is normally required to be at least two pipe sizes larger than the water service pipe.

Rolling a pipe out in a ditch to get water to a building isn't a complicated job, but there are some aspects of the job worth mentioning. For instance, the pipe must be installed at a depth where danger of the pipe's contents freezing is eliminated. This depth is known as the frostline, and it varies from geographical location to geographical location. For example, the frostline in Virginia may be 18 inches deep, and the same frostline in Maine may be 4 feet deep. Due to the differences in winter temperatures, the frostline is determined by local authorities.

Sizing

Before you can install a water service, you must know what size pipe is required for the job. In commercial work, the required size of the water service will usually be detailed on the blueprints. In residential work, the sizing of the water service is normally the responsibility of the plumbing contractor.

The water service for an average house will usually have a diameter of either ¾ or 1 inch. In cases where the water service is coming from a well pump, the pipe will often have a diameter of 1¼ inches.

The size of a water service pipe is determined by the linear distance the pipe runs and the number of fixture units it will provide water to. A fixture unit is a rating given to various plumbing fixtures by the local plumbing code. For example, in residential plumbing, a kitchen sink is rated as two fixture units, a bathtub carries a rating of two fixture units, a hose bibb is rated for three fixture units, a shower is rated for two fixture units, a flush-tank toilet is rated for three fixture units, a clothes washer hookup is rated as two fixture units, and a lavatory is rated at one fixture unit.

The fixture units given are common, but your local code may differ slightly. Keep in mind as you see specific numbers and examples throughout this book that you should always confirm local regulations in your area.

Apprentices will not be doing on-the-job sizing, but it will be advantageous for you to understand why one house has a ¾-inch water service and the next house has a 1-inch water service. Let me show you with some examples how the sizing for a water service might be done.

When you are ready to size a water service, you will need to refer to your code book. There will be tables in the code book that guide you in sizing a water service. For example, you will find a table or text that assigns fixture-unit values to various fixtures, just as I did a moment ago. That will be your starting point.

Once you know the assigned fixture-unit values, identify and list all the plumbing fixtures that will be served by your water service. For our sizing example, let's use a sample house that contains common fixtures.

To size the water service for this house, we will first have to find the total number of fixture units in the home. The three toilets equal nine fixture units. Three lavatories equal three fixture units, and two hose bibbs add another six fixture units. So far we have 18 fixture units. The two bathtubs are rated for a total of four fixture units, the kitchen sink adds another two fixture units, and the laundry hookup adds yet another two fixture units. Our total number of fixture units is 26. We now have the first variable in our sizing equation.

The next variable we must know is the operating pressure the plumbing system will maintain. If we are dealing with a municipal water system, we might assume the water pressure will be controlled by a pressure-reducing valve. For the sake of our example, let's assume the water pressure will be set at 50 pounds per square inch. We now have the second variable in our equation.

Now we need to know how long our water service will be. Let's assume the length of the water service will be 75 feet. That is the last piece of the puzzle we need to size the pipe for the water service.

Knowing the three variables, all we have to do is turn to our code book and check the tables and text. There will be guidelines that show us what size diameter the water service must have.

In this example, if we were working with the plumbing code that is prevalent along the East Coast, we would have to use a pipe with a minimum diameter of 1 inch for our water service. In this particular example, the water meter we would be connecting to could be sized at ¾ inch.

These size determinations are made with the simple use of tables and text provided in the local plumbing code. If the length of the water service piping in our example were only 40 feet, the water service could have a ¾-inch diameter. Figuring pipe sizing for a water service is easy when you have the help of your local code book.

Connecting to the water source

Connecting to the water source with a water service pipe can be done in many ways. We have already talked about water services used with well systems. When you are working with a municipal system, you might be fortunate to have the municipal water department make your connection for you. There are times when all the plumbers have to do is get the water service to the point of connection, and the city employees will make the connection.

The worst-case scenario will be if you have to tap into a water main with a saddle tap. Saddle taps are similar to the saddle valves used to connect ice makers to plumbing pipes but on a much larger scale. A saddle tap is secured to a water main, usually with bands that wrap around the main water line. Then a hole is drilled in the center of the tap and the water service pipe is connected.

Most connections will be made at the yoke of a water meter. The city will make the connection from their water main to the yoke, and the plumber will connect the water service from the interior plumbing system to the other end of the yoke. This type of connection is usually made with a mechanical fitting and is no big deal.

Laying the pipe

Laying the pipe is an important part of installing a water service. Most water services are pipes made either of plastic or copper. Both these materials can be damaged with rough handling. The bottom of the trench where the water service is installed should not contain any rocks or sharp objects. Care must be taken not to kink the pipe when laying the water service along the ditch. When the ditch is filled in with dirt, the dirt should be clean and free of rocks and other debris that might damage the pipe.

If you want to install a water service in the same ditch that the sewer occupies, you must observe some special rules. The water service cannot lay beside the sewer. It must be on an elevated shelf of undisturbed or compacted dirt (Fig. 14.1). Typically, the bottom of the water service pipe must be at least 12 inches above the top of the sewer pipe.

As we have discussed earlier, the water service must be buried deep enough to avoid freezing. The depth of the frostline will be established by your local plumbing code.

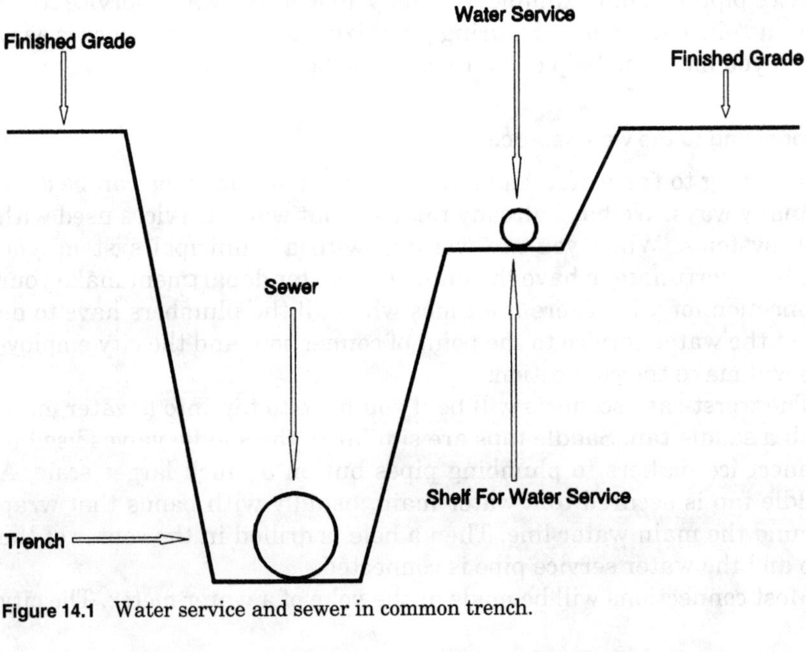

Figure 14.1 Water service and sewer in common trench.

Getting the pipe in a building

Getting the pipe in a building can be accomplished by going under the footing of the foundation or through the foundation. In either case, the water service pipe must pass through a sleeve when entering the building. The sleeve must normally be at least two pipe sizes larger than the water service pipe. The open space between the water service and sleeve must be filled to eliminate the risk of water or pest infiltration.

Once inside the building

Once inside the building, the water service may have to be converted to another type of pipe. If the material used for the water service is approved as a water distribution material, no conversion is required. However, if the water service is PE pipe, conversion must typically be made to a water distribution material within the first 5 feet of entry into the building.

Inspections

Water service piping requires a plumbing permit and inspection. With most pipes used in modern water services, there are no joints in the pipe. The pipe is either coiled copper or plastic and comes in long enough lengths to avoid coupling connections. Some inspectors will not require a pressure test of the pipe if no concealed joints will exist in the installation. But some inspectors will expect the water service pipe to be tested just the same as any other water pipe.

The test can be done with the use of water or air. The test pressure must be at least equal to the maximum operating pressure of the water that will flow through pipe.

Sizing Sewers

Sizing sewers is not a complicated process when you refer to the tables and text in your local plumbing code book (Tables 14.1 and 14.2). Residential sewers may not have diameters of less than 3 inches, and most will have 4-inch diameters. A 4-inch sewer for a house will carry a lot of fixture units, and there is almost never a need for a larger sewer on a residential home.

Sewers for commercial jobs can get much larger. Many commercial jobs will have 6-inch sewers, and a few will have 8-inch sewers. How will you know what size sewer pipe to install? You will refer to your local code book. There will be tables and text in it to guide you in sizing any common sewer. Let's use our same sample house from the water-service-sizing exercise and determine its sewer requirements. To re-

TABLE 14.1 Drainage Fixture-Unit Values

Fixture	Value
Bathtub	2
Bidet	1
Dishwasher	2
Kitchen sink	2
Laundry tub	2
Lavatory	1
Shower	2
Water closet (two pieces)	4
Water closet (one piece)	4

TABLE 14.2 Building Drains and Sewer Sizing*

Pipe size (in inches)	Number of fixture units allowed on pipe
2	21†
3	42‡
4	216

*This table is based on the drainage pipe falling ¼ inch for each foot it travels. For example, a 4-foot length of pipe would be 1 inch lower at the discharge end than at the inlet point.

†Any pipe with a toilet connected to it must have a minimum size of 3 inches.

‡No more than two toilets or bathroom groups may be on a 3-inch pipe.

fresh your memory, our sample house has the following plumbing fixtures: three toilets, three lavatories, two hose bibbs, two bathtubs, one kitchen sink, and one laundry hookup.

When we are dealing with waste pipes, these fixtures are still assigned fixture-unit ratings, but they are not the same ratings assigned for water-pipe purposes. You will find ratings that pertain to your area in a local code book. For this example, we will use common ratings for my area.

Each toilet gets a fixture-unit rating of four, and we have three of them; so we have twelve fixture units in the toilets. We have three lavatories, each with a rating of one fixture unit, for a total of three additional fixture units. There is no drainage rating for the hose bibbs, but the two bathtubs each carry a rating of two fixture units, for an additional four units. The kitchen sink is rated for two drainage fixture units. Our clothes washer drainage rating is three fixture units. This gives us a grand total of 24 drainage fixture units.

Now to size our sewer, we must know the number of toilets on the plumbing system, the amount of grade that will be placed on the sewer pipe, and the total number of drainage fixture units on the job. The total of the drainage fixture units is 24. We will grade the sewer with a standard grade of ¼ inch per foot, and we have three toilets on the system. We have all the information we need to use the tables and text in our code book for sizing the sewer.

If we refer to our code book, we might see that a 3-inch sewer, with a ¼ inch per foot of grade can handle 42 fixture units. This is certainly more fixture units than we have, so a 3-inch sewer looks good, until you read the fine print. A 3-inch sewer may not be burdened with the discharge of more than two water closets (toilets). Since we have three toilets, we cannot use a 3-inch pipe. We must move up to a 4-inch pipe. When we check the sizing table in our code, we might see that a 4-inch sewer can handle a total of 216 fixture units, with no restrictions on the number of water closets making up the total fixture count.

As you can see, a 4-inch sewer can carry a lot of fixture units. I can't think of a house that would require a sewer any larger than 4 inches.

Let's discuss the pros and cons of using a 3-inch sewer. If our sample house only had two toilets, we could run a 3-inch sewer. Would this be a wise move? Well, it would save a little money on the cost of materials, but it would limit the homeowner's future expansion possibilities. If the homeowner ever wanted to add a third toilet, the entire sewer would have to be replaced or bypassed. There are times, such as when bidding large jobs, that justify running 3-inch sewers, but on average, it is best to install a 4-inch sewer.

Installing Sewers

Installing sewers is pretty simple work. You get down in a ditch and run a pipe between two connection points. There are few fittings in the installation, and the job usually goes quickly. There are, however, some aspects of the work that require thought.

Most sewers will be connected to either a sewer lateral or a septic tank. In either case, the plumber should be connecting the sewer to an existing stub of a pipe. Since the pipe may not be of the same material used by the plumber, a universal coupling might need to be used. The couplings most often used are heavy rubber couplings that are held in place by two stainless steel clamps. These couplings can be used to join any common types of plumbing pipes together.

The first consideration in installing a sewer is the amount of fall, or

TABLE 14.3 Typical Minimum Drainage-Pipe Pitch

Pipe diameter (in inches)	Pitch (inches to the foot)
Under 3	¼
3 to 6	⅛
8 or larger	1⁄16

grade, the pipe will have on it (Table 14.3). Standard installations use a fall of ¼ inch per foot, but it is often acceptable to use a fall of either ⅛ or ½ inch per foot when installing sewers.

The determination of the amount of grade is important in making sure you can get from point A to point B without running out of grade. Sewers typically depend on gravity to drain, and if you don't have enough pitch on the pipe, the contents will not reach their intended destination. Let me give you an example.

Assume you are about to install a sewer. From the point where the sewer leaves the building to the point of connection with the city sewer lateral is a combined linear footage of 60 feet. When you shoot the grade of the two pipes with your transit, you see that the sewer lateral is 10 inches lower than your building sewer. If you don't have a transit, you could make this measurement with a string, a string level, and a tape measure.

Okay, you've got to cover 60 feet of ground, and you've got 10 inches of fall to work with. What grade will you put on your pipe? If you use a grade of ½ inch to the foot, by the time you have traveled the 60 feet, your sewer will be 30 inches lower than where it starts. This will put it about 20 inches below the connection point, and that won't work.

How about ¼ inch to the foot, will that work? No; at ¼ inch to the foot, your sewer will be 15 inches deep, a full 5 inches below the connection point. Will a grade of ⅛ inch to the foot work? Yes; you can drop the pipe with a grade of ⅛ inch per foot and be less than 8 inches in the ground at the end of your run. By choosing an average grade of ⅛ inch to the foot, you can make your connection without any trouble.

See why knowing your grade requirements is important? Are you wondering how we arrived at the numbers in the last paragraph? They are simple. We have 60 feet to travel. For a ½ inch grade, we know that for every foot the pipe travels it will drop ½ inch. So if it travels 60 feet, it must drop 30 inches. A grade of ¼ inch to the foot means the pipe will drop 1 inch for every 4 feet it travels; 4 goes into 60 fifteen times, so the pipe will be dropping 15 inches. This is simple math that helps you avoid a big problem. If your building sewer comes out lower than the main sewer connection, you are going to have to start all over again.

Okay, now you know what grade to use. Do you know how many cleanout fittings will be required along your way? Most codes require a cleanout within the first 5 feet of the sewer, as it leaves a building. Then, depending upon the code and the size of the sewer, additional cleanouts are required at 50- or 100-foot intervals. Usually, 3-inch sewers must have their cleanouts spaced no more than 50-feet apart. Sewers with diameters of at least 4 inches are often allowed a space of up to 100 feet between their cleanouts. If a change of direction is made with the sewer at an angle greater than 45 degrees, a cleanout is normally required.

Well, you are coming right along; you know what grade is required for your sewer, and you know where to put cleanouts. It is recommended that sewers not make short turns with the use of elbow fittings. Fittings that turn the piping at a 45-degree angle are considered to be the sharpest turns advisable. This is not to say an elbow fitting can never be used; it can, but it should be avoided whenever possible. When elbow fittings are used, they should be of the long-sweep variety.

Sewers should be deep enough in the ground that the dirt over them is at least 12 inches deep. The sewer pipe must be on firm, undisturbed or compacted ground. All sections of the sewer must be properly supported, and the installation of a sewer requires a permit and an inspection. Sewers can be tested with air or water, and the requirements of the test normally call for a 10-foot head of water or 5 pounds per square inch of air pressure.

Sewers, like water services, must be sleeved where they penetrate a foundation wall or under a footing. The pipe sleeve should be at least two pipe sizes larger than the sewer.

Special Advice

Most sewers today are made of ABS or PVC pipe. These types of pipe require solvent-weld joints, and the joints may be fouled by dirt or mud. It is important to keep the ends of your pipe and the hubs of your fittings dry and clean.

Trenches for sewers and water services can become death traps. Carefully inspect trenches before entering them, and always abide by good safety procedures.

Chapter

15

Underground Plumbing

Underground plumbing might be compared to the foundation of a home. The plumbing that runs underground and under concrete floors is the foundation of a plumbing system. Water service and sewer pipes are the simplest types of underground plumbing, but they are not the only types. They are, in fact, only a small percentage of the underground plumbing for many homes and commercial buildings.

Plumbing that is buried in the ground and under concrete floors is known as underground plumbing. Many plumbers refer to this plumbing as groundwork. The groundwork of a plumbing system is often the first phase of plumbing to be installed. If the groundwork plumbing is installed incorrectly, the entire plumbing design may have to be altered, or the groundwork might have to be exposed and rearranged. Breaking up a new concrete floor to move a pipe 1 inch to the left is not only embarrassing, it's expensive.

To an inexperienced eye, seeing pipes laid on the ground, waiting to be covered with concrete may provide a false impression as to the importance of installing underground plumbing properly. A finished groundwork system looks simple enough. There will be pipes laid in and on the ground, with vertical pipes turned up at various locations. To look at such a layout, the job of installing underground plumbing seems simple. However, if the pipes are not installed properly, the remainder of the plumbing in the structure may never work properly.

Apprentices who are involved in construction plumbing often get their first taste of the trade by working with underground plumbing. The apprentices are introduced to shovels and jackhammers and shown how to prepare ditches for the pipes to be installed. They may help pull strings and take measurements prior to the installation of the pipes. Cutting pipe and delivering fittings to the plumber will be only one part of the apprentice's job during the groundwork installation. When all of the pipes are installed, the apprentice will help test the installation. Depending upon the circumstances, it will not be unusual for the trainees to haul buckets full of gravel to bed the pipe in. Underground plumbing not only requires precise measurements, the work can be physically demanding.

This chapter is going to provide insight into the makings of an underground plumbing system. You will learn, from start to finish, what to expect when assisting with your first groundwork installation. In addition, you will be taught many typical code requirements for underground plumbing. The rules for underground plumbing are a little different from those used for above-ground plumbing. Now, let's get started.

Our sample job involves installing the underground plumbing for a block of townhouses. There are six townhouses in the block, separated by common party walls. These townhouses will have three finished levels of living space, one of which is the walk-out basement. Each townhouse will contain two complete bathrooms and a laundry hookup on the top floor, one half-bath and kitchen on the mid-level, and a rough-in for a full bath on the lower level. None of the rough plumbing may be left exposed when the houses are complete.

The main foundation, one large rectangle, for the townhouses has been installed. However, the party walls have not yet been constructed. The ground within the foundation is undisturbed earth, and there are piles of gravel at each end of the foundation. The gravel will be distributed over the inner area of the foundation at a later date. Now that you have an overview of the job site, let's unload the truck and get to work.

You are a new apprentice, with limited experience. I am your plumber. After arriving at the job, I ask you to unload the tools and materials from the truck. You unload the boxes of plastic drainage fittings and the lengths of pipe. Afterward, you unload the tools: shovel, pick, grade level, torpedo level, line level, hammer, tape measures, saw, and digging bar. The rest of what we need is unloaded next: cleaner, primer, glue, string, flour, and rags.

While you have been unloading the truck, I have had the blueprints for the job spread across the hood of the truck, studying them. I've verified the depth of the main sewer and decided on how the underground plumbing will be installed.

We take the big tape measure and check the measurements on the main foundation. Sometimes the foundations are not made to the exact specifications shown on the blueprints, and that can require making several alterations for interior walls and plumbing. The foundation checks out, and we begin pulling strings to simulate the walls that will be built later.

After the string walls are all in place, I show you where the sewer and water service will enter the first townhouse. We will need holes dug under the footing of the foundation for these pipes to enter through. You start by digging a hole on each side of the foundation wall. When the hole is large enough and deep enough, you use the digging bar to tunnel under the footing.

While you are digging, I'm spreading flour inside the foundation to mark the locations of trenches. Once the lines are laid out, I sort through the box of fittings and lay them at appropriate places along the plumbing path. Next, I cut short pieces of pipe, two pipe sizes larger than the pipes that will go through them, to be used as sleeves for the water service and sewer.

When you have finished tunneling the holes, you start digging along the lines of flour. You start at the holes under the footing. We will use two 45-degree offsets to come up from the deep hole to a more shallow grade. You grade the ditches with a ¼-inch-to-the-foot grade. I've shown you how to lay a piece of pipe in the ditch and check it with the grade level to maintain a consistent grade on the trench.

While you are digging the ditches in the first foundation, I step across to the next townhouse, still in the same main foundation, and begin laying out its plumbing. I remind you not to hit our string walls while doing your work.

Once the ditches are dug, I begin to install the plumbing, with your help. I put the sleeve for the sewer in the hole under the footing. You cut and hand me a 5-foot length of 4-inch pipe. I glue a test cap into the end of the pipe and slide it through the sleeve. A 45-degree fitting is installed on the other end of the pipe. That fitting points upward. Taking another 45-degree fitting, I hold it in line with our ditch and give you a measurement for a piece of pipe.

You cut the pipe, scrape off the burrs of plastic, and hand it to me. I apply the cleaner, primer, and glue and then install the pipe between the two 45-degree fittings. By creating this offset, we have avoided digging deep ditches throughout the foundation.

The next fitting installed is a combination wye and eighth bend. This long-sweep fitting allows us to turn up a pipe for a cleanout. We go about our business installing the plumbing until you hand me a short-turn elbow. I remind you that short-turn fittings are not permitted for use under concrete. When we are getting ready to rough-in the drains for the downstairs bathtub and lavatory, you ask why we are using 2-inch pipe, instead of 1½-inch pipe. I explain that 1½-inch pipe would be adequate if the drains were above ground but that the smallest pipe allowed for underground drainage is 2-inch pipe.

As we go along, we check our pipe grade with the grade level frequently. When we turn up a long-sweep elbow for a stack, we check it with our torpedo level to make sure the fitting is level. If the fitting is not level, the stack will be crooked when it comes out of the floor.

We protect our underground trap for a bathtub with a tub box (Fig. 15.1). The box will keep concrete away from the trap and allow us to connect to it after the concrete floor has been poured.

The pipe we turn up for the basement toilet will be covered with a special spacer. The plastic cap (Fig. 15.2) will keep concrete from surrounding the pipe and prohibiting us from installing a closet flange at

Figure 15.1 Typical tub box for underground trap.

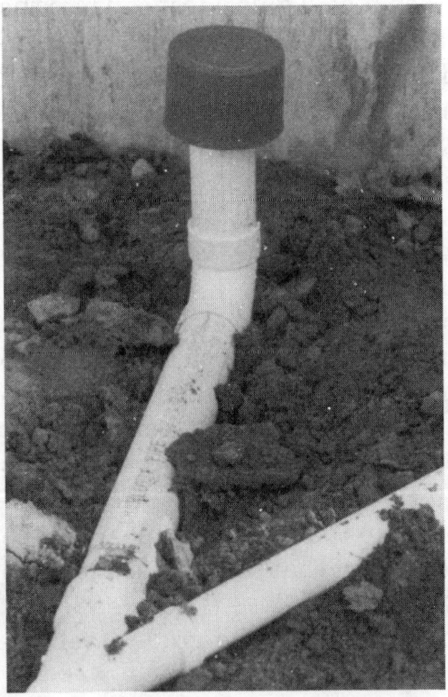

Figure 15.2 Spacer-cap on a water-closet stub-up.

Figure 15.3 Measuring the rough-in of a water closet.

a later time. Before installing the cap, we check the position of our pipe to make sure it is roughed-in properly (Fig. 15.3).

During our work we are very careful not to knock our string walls out of alignment. The pipes are installed with precision, in and around the strings (Fig. 15.4). When all of the pipes are installed, you take short lengths of ½-inch rigid copper tubing and drive it into the ground on each side of the fittings holding our stacks (Fig. 15.5). The copper stakes will not attract termites, and they will keep our pipes secure in the locations we have picked for them.

Once the pipes are staked into position, you haul gravel in a 5-gallon bucket and use it to fill in voids under fittings and around the pipes. When the plumbing is finished, we check all of our wall and stack measurements again. If anything has been installed improperly, we want to correct it now, not later.

Our last step in piping is the installation of the water service. We uncoil a roll of soft copper tubing and run it from the entry hole to the proposed location of the mechanical room. You drive a copper stake in the ground and secure the turned-up water service to the stake with duct tape (Fig. 15.6). A piece of pipe insulation is slid over the water service where it turns up out of the concrete floor. The foam insulation will protect the copper pipe from being damaged by the concrete. Without the insulation, the copper might vibrate in the concrete and wear a hole in itself. It is also possible that a chemical reaction might occur if the copper were left in direct contact with the concrete.

Figure 15.4 A groundwork stack turned up between strings.

When we are satisfied that the pipes are in the right places, you go around and install test caps on all the pipes except for one. We have one pipe that stands 10 feet above the groundwork. This is our test pipe. You will go over and dig out the ditches in the next townhouse while we wait for our joints to dry.

When we get back to the shop, we call for a plumbing inspection on the underground plumbing for the following day. Before the inspector arrives, you fill the groundwork with water until it is running out of the 10-foot pipe. The inspector comes and approves the installation. We drain the water out of the pipes and move over to the next townhouse to continue our plumbing.

This has been a pretty fair example of how an underground plumbing job might be done. Now that you have an idea of what the basics are for installing groundwork plumbing, let's take a look at some specifics you will find useful.

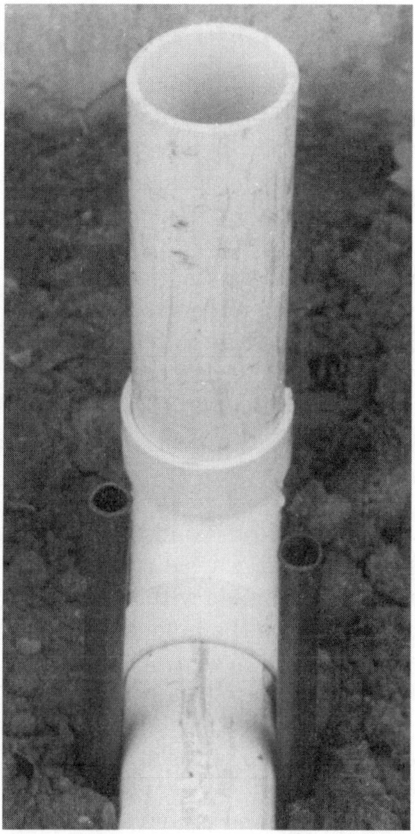

Figure 15.5 Example of staking underground pipes into place.

Figure 15.6 Example of a water service staked into place.

Tools

Most all the tools needed to install groundwork plumbing were mentioned in the example. As an apprentice, you are likely to do a lot of digging and manual labor. A tough pair of gloves will save your hands from a lot of abuse. Knee pads can come in handy for this kind of work, too. Eye protection should be worn if you are making a hole in a foundation or using a pick on rocky terrain.

Every apprentice should have a tape measure and know how to read it. Measurements are critical when working with underground plumbing and turning up stacks and vents. If you don't already know what each of the marks on a tape measure mean, find out. If a pipe winds up outside of a wall, somebody is going to be in trouble.

Pipe Preparation

Installing underground plumbing sometimes requires additional pipe preparation. The working conditions are almost always dirty, and they can be downright muddy. Good joints cannot be made with plastic pipe and fittings that are wet, dirty, or muddy. Since the pipe is often laying in the dirt prior to use, it is the helper's job to clean it before the plumber gets it, not necessarily with a liquid pipe cleaner but at least with a clean, dry rag.

After plastic pipe is cut, there are often ragged burrs of plastic left around the cut. The shreds of plastic should be removed. Some plumbers use a knife to trim away the plastic and others wipe them away with a rag.

Cutting the pipe squarely is important in all aspects of plumbing, and it can be hard to do when the cutting hands are inexperienced. Many plumbers provide their apprentices with a miter box to hold the pipe for a straight cut. Experienced apprentices and plumbers don't need the assistance of a cutting guide. They can pick the pipe up, hold it against one leg, and saw it off. However, you must get good, even cuts, and if you need a miter box to do it, get a miter box.

If the ends of the pipe are not cut squarely, the joint made with the pipe may leak. When one side of the pipe is longer than the other, the hub of the fitting is not filled with pipe. The long end will butt against the back of the hub, and the shorter end will not be seated completely. This is a problem many new apprentices face, but a miter box solves the problem.

If you favor cutting your pipe with a hacksaw, you probably won't be able to cut squarely pipe that has a diameter of more than 3 inches. The larger pipes are difficult to cut straight with a hacksaw. A universal saw or a standard carpentry saw will get the job done on larger pipes.

Most codes recommend, if not insist, that PVC pipe be cleaned with an approved cleaner, primed with an approved primer, and glued with an approved solvent. Many plumbers skip the cleaning and priming, but they shouldn't. PVC can be finicky, and it will leak if the joints are not prepared and made properly.

ABS pipe is much less temperamental than PVC when it comes to substandard working conditions. A little dirt in an ABS fitting probably won't cause a leak. However, ABS should also be clean and dry before making a joint.

Digging the Ditches

When you are digging the ditches for underground plumbing, avoid digging them too deep. If the trench is not dug to grade, there will be voids under the pipes. These voids must be filled, usually with crushed stone, and guess who gets to haul it in with 5-gallon buckets? Taking care in the preparation of the ditch will save work and time later.

Watch Out for the Strings

While you are working around the groundwork, watch out for the strings. Tripping over the strings that simulate wall locations is a great way to get your plumber angry with you. Other common causes for string disruption include hitting it with a piece of pipe, letting a shovel handle fall on it, and tossing dirt onto it.

Securing the Pipes

Securing the pipes is an important part of installing underground plumbing. Other trades will be working around the pipes, and if the pipes aren't secure, they can be knocked out of place. When this happens, the plumber usually gets the blame. General contractors don't run out and yell at the concrete finishers when a pipe comes out in the wrong place; they look for the plumber.

The pipes should be laying on firm ground or a suitable bed of stone. Stone should be laid alongside the pipe to keep it in position. Metal stakes should be installed at all turn-ups and at various points along runs of pipe to keep the system from being moved. Make sure the stakes are short enough not to be left sticking out of the concrete when it is poured.

Judging Floor Height

Judging the floor height of a finished floor can be accomplished with a string and a line level. The job-site superintendent will supply a grade mark for the finished floor level. This will often be done with a mark on a foundation wall, but it may be done with a grade stick. Once you have the finished grade elevation, you can pull a string across your work area and adjust its height with the line level. This will keep your stakes and fittings from sticking up out of the concrete floor.

Cap All Pipes

After your piping has passed inspection and been drained of its air or water, cap all pipes. Never leave an open pipe in a groundwork installation. Otherwise, objects will be dropped in the pipes that may clog the system.

Why Do You Have to Be So Careful with Underground Pipes?

There are many reasons for using extra care with the placement and confinement of underground plumbing. Much of the underground plumbing installed is covered with stone, reinforcing wire, and concrete. Rebar is sometimes added for extra strength in the floor. If the underground plumbing becomes damaged or is in the wrong place, correcting the problem can get very expensive. To impress upon you the importance of these matters, allow me to give you some real-world examples of what could happen if you don't take care of your underground plumbing.

A rock clog

This first example deals with a rock clog. Let's say you have finished your underground plumbing, gotten your inspection, and left the job. Being in a hurry, you neglected to cap all of the stacks and vents. Over the weekend, some kids come on the job site and are playing around the plumbing. One of the children decides it would be great fun to play a rock-toss game.

The kids stand back from the plumbing and take turns throwing rocks at the open end of one of your plumbing stacks. The games goes on for a while and a number of rocks are tossed into the pipe; some of them are large. Tiring of the game, the children move on to other activities.

No one, except the kids, knows the rocks are in the pipe. The concrete company does its job and pours and finishes the concrete. As the structure is built, you return and rough-in the above ground plumbing. Nothing appears to be wrong with the plumbing. When the final plumbing is complete and you test the fixtures, everything seems to work fine. Since your test is done only with water, the rocks don't cause much of a problem.

After the building begins to be used and toilets get flushed, a problem develops. The waste from the toilets is slow in going down one of the drains. In time, the pipe becomes blocked solidly. An electric drain cleaner is employed to clear the stoppage, but it can't. It's cutting head can't get past the large rocks. Now what do you suppose will have to happen?

Since the plumbing is under warranty, the plumbing contractor will have to break open the concrete floor and correct the problem. It will turn out that the rocks are clogging the pipes, and the pipe will have to be cut and repaired. Then the concrete floor will have to be repaired. Who is going to pay for all of this? The plumbing contractor is. Who is going to be blamed for the problem? The general contractor will blame the plumbing contractor. The plumbing contractor will blame the field plumber, and the plumber will point a finger at you, the apprentice. All of this could have been avoided if the ends of the pipes had been sealed.

The unsecured stack

This is the story of the unsecured stack. The underground plumbing has been installed, but you forgot to secure one of the stacks. When the person responsible for spreading and grading the gravel in the foundation hole bumps into the stack, it moves several inches. The well-meaning worker moves the pipe back to where it looks like it should be and continues to grade out the foundation.

The concrete people come in and do their thing. When the carpenters begin framing their interior partitions, they find that the pipe that was hit sticks out of the wall by more than an inch. The wall cannot be moved, so the pipe must be relocated. We are back to having to break concrete, move a pipe, and so on. The chain of blame will be the same as in the earlier example. This same type of problem could arise if a measurement was not made accurately.

These examples are very real. I've seen them both happen. Having to break up a new concrete floor and alter work you have already finished is no fun, and if it happens often, you won't have a job for long.

Good Installation Practices

There are many good installation practices that should be observed when installing underground plumbing. Some of them are mandated by the plumbing code, and some are just plain common sense. Let's take a moment to look at some of the work habits you should keep in mind when working with underground plumbing.

Pipe size

The minimum pipe size for drainage pipe under concrete should be 2-inch pipe. This includes traps, such as for bathtubs and floor drains, when the traps are under the concrete. Since the pipe is not easily accessible, the larger size makes drain cleaning easier and more effective.

Long-sweep fittings

Long-sweep fittings are the only type of drainage fittings that should be installed under concrete. Use of short-turn elbows and other short-radius fittings can result in clogged drains.

Sleeves

Any time a pipe is passing under or through a foundation wall, it should be run through a sleeve that is at least two pipe sizes larger than the pipe. These sleeves should be filled with an approved compound to keep water and rodents out of the building.

When copper pipe and tubing is turned up to have concrete poured around it, the copper should be protected in a way that it will not come into direct contact with the concrete. Abrasion and chemical reactions may occur if the pipe is not properly protected.

Water services

Water services should be rated with a minimum working pressure of 160 pounds per square inch at 73.4 degrees F. If the water pressure in the pipe is expected to exceed 160 pounds per square inch, a pipe with a higher pressure rating should be used.

Cleanouts

Cleanouts should be installed at the base of each soil stack. Plugs for cleanouts should be made of brass or plastic. The cleanout plugs should have raised heads, unless they will cause a safety hazard, in which case recessed heads should be used. A cleanout should be installed at the junction of the building drain and the building sewer.

These are not all of the code requirements for underground plumbing. However, this list of pointers will help you with the installation of groundwork plumbing. Refer to your local code book for all applicable code restrictions and regulations.

Things to Remember

Remember that the underground plumbing is the foundation of your plumbing system. If the pipes in your groundwork do not match up with the needs of your above-ground plumbing, you are going to be in for trouble.

Can you imagine how you would feel if the floor drain you roughed-in for the relief-valve discharge of a water heater wound up outside of the mechanical room? How would you feel if you were roughing-in a soil stack for a townhouse and turned it up on the wrong side of the party wall? Your pipe would be in the wrong house; what would you do?

Underground plumbing may not look like much laying in the ditch, but it is critical to the successful operation of the rest of the plumbing system. Don't get sloppy when roughing-in the groundwork. Now, let's move ahead to the next chapter and study drainage systems.

Chapter

16

Drainage Systems

Drainage systems are a part of every plumbing system. The water comes in through water distribution pipes, and it goes out through the drainage system. You saw a little about drainage systems in the last chapter. However, above-ground drainage systems can be somewhat different from underground systems.

This chapter is going to deal with common above-ground drainage systems. You will learn the difference between a building drain and a building sewer. Every good apprentice needs to know how far apart to space pipe hangers, and this chapter will show you.

We will talk about combination waste and vent systems. Drainage fixture units will be discussed, and you will learn how to size drainage pipes. By the time you finish this chapter, you will be well prepared to help with the installation of a drainage system. You will know which fittings can be used for what purposes, and you will know what must be done to protect your pipes.

Now, let's set the stage with a simulated job experience. Do you remember the townhouse in which we installed the underground plumbing? Good, because that's the house we are going to install our make-believe drainage system in. Now, if you are ready, we will get to work.

You will unload the truck while I lay out the drainage system. We will need the following tools: right-angle drill, with bits; reciprocating saw, with blades; extension cord; stepladders; tape measures; torpedo levels; hammer; and metal-cutting snips. We will, of course, need our pipe, fittings, hanging strap, stove bolts, glue, primer, cleaner, rags, and roofing nails.

By the time you get the truck unloaded, I'm well on my way to getting all the holes we will need marked out on the studs and joists. You plug in the extension cord, put on your safety gear, and get ready to drill and cut. Once I'm far enough out in front of you, you begin drilling and cutting the holes I have marked.

By the time the holes are all cut, your arms are a little weary and you're covered in sawdust. You've been smart and fortunate; your drill bit didn't bite into any nails or knots in the wood.

Now that the holes have been opened up, we are ready to start running pipe. I could have you drill out the holes for the water pipe while I install the drainage, but I want to show you how the job is done, so you get to help with the drainage.

We start our work at the beginning of one of our stacks. This house has two main soil stacks: one for the powder room and one for the two upstairs bathrooms. The first thing we do is install a test-tee fitting. The test-tee fitting will be used when we test our work, but it will also serve as a cleanout. Coming out of the top of the test tee, we run our waste and vent pipes.

The cutting and gluing is the same as when we did the groundwork plumbing. However, we are able to use some short-turn fittings and some 1½ inch pipe. The grade on our pipes is a constant ¼ inch to the foot. We don't need a grade level to check our pipe pitch; we are able to confirm the grade with our tape measures.

When the pipe is being hung under the floor joists, we can measure from the top of the pipe to the bottom of the joists to check our grade. For example, if we are installing an 8-foot piece of pipe, the downhill end of the pipe should be 2 inches lower than the higher end. If it is, we have ¼ inch to the foot of fall. We don't want to make the grade too great or there might be drainage problems. When the pipe falls horizontally too fast, waste water can rush down the pipe and leave toilet paper and such stranded, high and dry. This can cause stoppages in the pipe.

As we work along, you install hangers on the pipe at 4-foot intervals. This spacing is determined by tables and text in the code book (Tables 16.1 and 16.2). The work progresses well, and by the end of the day, most of the drainage work is done. You will still have to install nail plates (Fig. 16.1) to protect some of the pipes that are at risk of being hit by a drywall screw, but all in all, the job has gone well.

You load the truck while I inventory the materials used on the job and plan a list of material for the next day. On the way back to the shop, you ask me many questions about why we did what we did. Wanting to help you accelerate your learning process, I explain to you why the job was done in the manner you observed.

Well, there you have a typical day of installing a drainage system. It has been abbreviated a bit to leave space in the chapter for specifics, but you now have a good idea of what it will be like to work with an

TABLE 16.1 Common Vertical Pipe-Support Intervals

Type of vent pipe	Maximum distance between supports, in feet
Lead pipe	4
Cast iron	15
Galvanized	15
Copper tubing	10
ABS	4
PVC	4
Brass	10
PB pipe	4
Aluminum	15

TABLE 16.2 Common Horizontal Pipe-Support Intervals

Type of vent pipe	Maximum distance between supports, in feet
PB pipe	32 inches
Lead pipe	Continuous
Cast iron	5 or at each joint
Galvanized	12
Copper tube (1¼ inch)	6
Copper tube (1½ inch & larger)	10
ABS	4
PVC	4
Brass	10
Aluminum	10

Figure 16.1 Nail plate.

above-ground drainage system. Now, let's listen in on the answers to those questions you asked in the truck.

What Types of Pipe Are Allowed for Above-Ground Drainage?

Every local code can have its own version of what is and is not an approved material for above-ground drainage, but some of the most commonly approved materials are PVC, ABS, copper, cast iron, galvanized steel, brass, and even lead and aluminum.

The most common materials used in modern drainage systems are PVC and ABS. However, copper and cast iron are still used for some special applications. Galvanized pipe has been used for small drains, but it has a tendency to rust and clog up, so it's not used much any more.

Is the Spacing of Hangers the Same for All Drainage Pipes?

The spacing varies for the placement of hangers on horizontal and vertical drains; it depends on the type of pipe being installed. ABS and PVC pipe is normally supported horizontally and vertically at 4-foot intervals. Cast-iron pipe usually has a horizontal support at every joint or at intervals of 5 feet, whichever is closest. But vertical sections of cast iron are generally supported at 15-foot intervals. Galvanized pipe is supported at 12-foot intervals horizontally and 15-foot intervals vertically. Copper pipe with a diameter of 1¼ inch or less has horizontal supports spaced at 6-foot intervals and vertical supports at 10-foot intervals. Larger-diameter copper is supported at 12-foot intervals horizontally and 10-foot intervals vertically.

All of this spacing information can be found in your code book, and different codes allow and disallow different procedures and materials. To be safe, you must always check your local code requirements before installing plumbing.

How Do You Know What Size Pipe to Install?

Your code book will provide you with the requirements for minimum pipe sizing. You will be shown, either in a table or in the text, what the smallest trap you can use for a specific fixture is (Table 16.3). Ratings for drainage fixture units will also be given to help you size drains that

TABLE 16.3 Recommended Trap Sizes

Type of fixture	Trap size, in inches
Bathtub	1½
Shower	2
Residential toilet	Integral
Lavatory	1¼
Bidet	1¼
Laundry tub	1½
Washing machine standpipe	2
Floor drain	2
Kitchen sink	1½
Dishwasher	1½
Drinking fountain	1¼
Public toilet	Integral
Urinal	2

serve more than one fixture. In addition, you can never decrease the size of a drain on the way to the sewer. This means that if you start out with a 2-inch trap, you cannot reduce the downstream side of the trap to a smaller pipe size. You can run the 2-inch drain into a 3-inch drain but not into an 1½-inch drain.

Now, as I was saying about the tables and text in your code book, you can check those sources for guidance in sizing your drains. Let me give you a few examples.

Let's say you need to pipe the drainage for a washing machine. What size drain will you use? If you look in your code book, you will find the required size of the trap. For example, the code might tell you that the trap for a clothes washer must be at least 2 inches in diameter. Once you know this, you know to run a 2-inch drain for the fixture. A bathtub will probably require a 1½-inch trap, unless you are installing it underground; then it would have to be a 2-inch trap. A lavatory will normally be allowed to have a 1¼-inch trap. The list goes on, and if you look in your code book, you can find the information you need.

Close by the section in your code book on trap sizes will be ratings for fixture units. Each common plumbing fixture will be assigned a rating. For example, a clothes washer might be rated at three fixture units, a bathtub at two, a dishwasher at two, a lavatory at one, and so on. Once you have found this information in your code book, you are well on your way to being able to size a drainage system. There will also be a table or text to allow you to pick a trap size based on the number of fixture units it will serve (Table 16.4).

TABLE 16.4 Common Fixture-Unit Requirements on Trap Sizes

Trap sizes, in inches	No. of fixture units
1¼	1
1½	2
2	3
3	5
4	6

How Do I Size a Horizontal Fixture Branch or Stack?

You size a horizontal fixture branch or stack much like we sized sewers a few chapters back. You will find a table or text in your local code book that will provide the criteria for sizing these pipes (Table 16.5). Let me give you an example of some common code requirements for this type of work.

Assume you have a 2-inch drain that is running horizontally. By checking with a table or the text in your code book, you find that the total number of fixture units a 2-inch pipe being used as a horizontal branch can carry is six. All you have to do is refer to the fixture-unit ratings for various fixtures, and you can quickly determine how many of the various fixtures can be carried by the one pipe.

When you are working with toilets, remember that the drain for a toilet must have a diameter of at least 3 inches. Also keep in mind that even if a 3-inch pipe is rated for enough fixture units to carry more than two toilets, it is rare for more than two toilets to be allowed on a 3-inch pipe. Just because the pipe is rated for enough fixture units doesn't mean you can pipe as many toilets to the pipe as the fixture-unit method would allow. Normally, toilets are the only fixtures that are exceptions to the fixture rating requirements.

TABLE 16.5 A Typical Stack-Sizing Table

Pipe size, in inches	Fixture-unit discharge on stack from a branch	Total fixture-units allowed on stack
1½	2	4
2	6	10
3	20*	48*
4	90	240

*No more than two toilets may be placed on a 3-inch branch, and no more than six toilets may be connected to a 3-inch stack.

If you are interested in establishing how many fixture units a 2-inch pipe being used for a stack can carry, you use similar sizing methods. For example a 2-inch stack with three or fewer branch intervals can accommodate up to ten fixture units. A branch interval is where a horizontal drain connects to a stack. If the stack is tall enough, having more than three branch intervals, it can carry up to 24 fixture units. Again, keep in mind that not all codes are the same. Check your local code requirements before installing any plumbing.

Do All Drains Have to Have a ¼-inch-to-the-Foot Grade?

The grade required for a drain normally depends on the size of the pipe and where it is being installed. For example, when you are working with standard, above-grade drainage systems, you will probably have three increments of pipe pitch to choose from. Pipes with diameters of 2½ inches or less will be required to fall at a rate of ¼ inch to the foot. Pipes with diameters of between 3 and 6 inches can be installed with only ⅛ inch to the foot of fall. The pipe pitch for pipes with 8-inch diameters and larger might only be 1⁄16 inch to the foot.

What Fittings Can Be Used to Change Direction?

The fittings that can be used to change direction depend on whether the change in direction is horizontal to vertical, vertical to horizontal, or horizontal to horizontal. The change of direction makes a difference in the fittings approved for the purpose. I will give you some common examples of approved fittings for each type of change in direction (Table 16.6), but be advised: There are usually some underlying exceptions to these choices.

Horizontal to horizontal

The fittings commonly approved for changing direction in a horizontal-to-horizontal method include sixteenth bends, eighth bends, sixth bends, long-sweep fittings, wyes, and combination wye and eighth bends.

Horizontal to vertical

To change direction from horizontal to vertical, you might use any of the following fittings: sixteenth bend, eighth bend, sixth bend, long-sweep fittings, wyes, combination wye and eighth bends, short-turn

TABLE 16.6 Fittings for Changes in Direction of Drain Waste and Vent Pipes*

Type of fitting	Horizontal to horizontal	Vertical to horizontal	Horizontal to horizontal
Sixteenth bend	Yes	Yes	Yes
Eighth bend	Yes	Yes	Yes
Sixth bend	Yes	Yes	Yes
Quarter bend	Yes	No	No
Short sweep	Yes	Yes	No
Long sweep	Yes	Yes	Yes
Sanitary tee	Yes	No	No
Wye	Yes	Yes	Yes
Combination wye and eighth bend	Yes	Yes	Yes

*Uses of fittings that are allowed are labeled Yes and uses of fittings that are not allowed are labeled No.

quarter bends, and sanitary tees. Remember that we are talking about above-ground drainage systems. The rules for underground systems vary slightly.

Vertical to horizontal

A change in direction from vertical to horizontal can usually be made with any of the following fittings: sixteenth bend, eighth bend, sixth bend, long-sweep fittings, wyes, and combination wye and eighth bends.

What Do I Need to Know about Cleanouts?

The horizontal drains in a building with diameters of 4 inches or less must be equipped with cleanouts at intervals not to exceed 50 feet. If the drain diameter is larger than 4 inches, you can space the cleanouts 100 feet apart.

There must be a cleanout installed where the building drain meets the building sewer. If the cleanout is for an underground drain, the pipe with the cleanout fitting on it must rise to the finished grade of the ground or floor.

Cleanouts are typically the same size as the pipe they serve. However, with pipes having diameters of more than 4 inches, 4-inch cleanouts are acceptable.

Cleanouts must be installed with the flow of the drain. In other words, if you feed a drain-cleaning cable into the cleanout, it must naturally go with the flow. If the cleanout fitting doesn't have a throat, like a test tee, the cleanout can be installed so that the access to the drain is at a 90-degree angle. The point is, don't install a sanitary tee upside down for a cleanout.

You cannot allow the cleanout plug to be concealed to a point where it is not accessible. It would typically be a code violation to hang drywall over a cleanout plug. You could allow the installation of a hinged door over the plug, but the plug must be accessible.

If a horizontal drain takes a turn at more than a 45-degree angle, a cleanout must be installed at the point of the change in direction.

How Much Open Space Do I Have to Have for a Cleanout?

If the pipe you are working with has a diameter of at least 3 inches, you must allow 18 inches of clear space in front of the cleanout. On smaller pipes, a clear space of 12 inches is required. This is a requirement that many plumbers overlook.

There have been many times when I have seen cleanouts in inaccessible locations. They get turned up toward the floor joists, they are installed sideways, facing a wall stud, and in a number of other inappropriate ways.

How Do You Know How Tall to Make the Laundry Standpipe?

Your code book will tell you how tall to make the laundry standpipe. One common rule pertaining to these indirect wastes is that the standpipe must be at least 18 inches tall but may never be more than 30 inches tall (Fig. 16.2). Anything between those two sizes is often acceptable.

Is It Possible to Pump Sewage Up to a Drain?

It is possible to pump sewage up to a drain, but it requires a special pump and installation arrangement. Let me explain how we could have installed a pump system for the basement bathroom we roughed-in at the townhouse.

The building drains for houses with basements often go out of the foundation wall at between 4 and 5 feet above the basement floor. This makes it impossible to install a bathroom in the basement with conventional methods. Under these conditions, the drainage from the basement bathroom must be pumped up to the building drain.

An interior sewage pump is not complicated to install. The first step

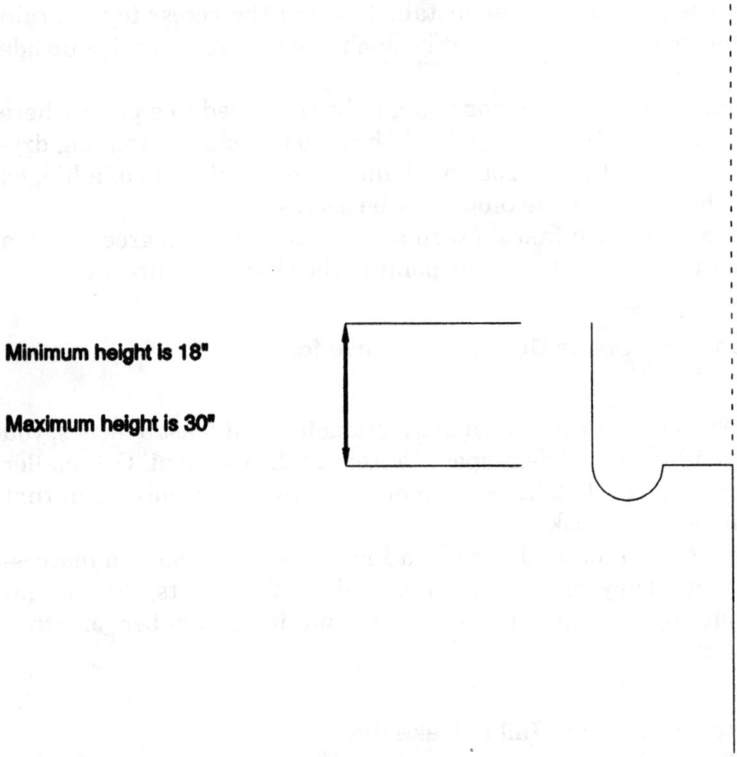

Figure 16.2 Washing-machine standpipe height.

is burying a special sewage sump in the floor of the basement. The sump will have a 4-inch hole in the side of it, up near the top. This is where the main drain from the bathroom will attach. If the drain coming from the bathroom is a 3-inch pipe, a 4- by 3-inch reducer can be used to adapt the smaller pipe to the larger opening.

Remember, drains can always get bigger as they flow to the sewer, but they may never get smaller. The only exception is a 4- by 3-inch closet bend. This is a special elbow used to turn up from a 3-inch pipe to a toilet with a 4-inch flange. These fittings are rarely used anymore, and they are the only exception to the sizing rule.

The 4-inch pipe that goes into the socket of the sump will normally be installed with the use of a rubber, donut-like adapter. Once the underground plumbing is installed and connected to the sump, all drainage from the plumbing fixtures in the basement bathroom will run to the sump.

This sump is not small. Sumps vary in size, but they are generally a few feet deep and up to 30 inches across at their top. The top for the sump will have two 2-inch holes in it, and there will be a smaller hole for the electrical cord from the pump.

A special sewer pump will be set into the sump. These pumps resemble sump pumps, but they are not the same thing (Fig. 16.3). The pump will be sized based on the vertical distance it will have to lift the waste.

Pump Model No.	Solids	Order No.	Control Switch①	Pump Control (NEMA1)
3871	¾"	EPO411A	NR	NR
		EPO411CA	NR	NR
		EPO411	A2-9	NR
		EPO412	A2-6	NR
3885	¾"	WE0311L	A2-9	NR
		WE0312L	A2-6	NR
		WE0311M	A2-9	NR
		WE0312M	A2-6	NR
		WE0511H	A2-5	NR
		WE0512H	A2-6	NR
		WE0532H	A2-7	A3-5034
		WE0534H	A2-7	A3-5034
		WE0511HH	A2-5	NR
		WE0512HH	A2-6	NR
		WE0532HH	A2-7	A3-5034
		WE0534HH	A2-7	A3-5034
		WE0712H	A2-7	A3-2012
		WE0732H	A2-7	A3-5034
		WE0734H	A2-7	A3-5034
		WE1012H	A2-7	A3-2012
		WE1032H	A2-7	A3-5034
		WE1034H	A2-7	A3-5034
		WE1512H	A2-7	A3-2012
		WE1532H	A2-7	A3-5034
		WE1534H	A2-7	A3-5034
		WE1512HH	A2-7	A3-2012
		WE1532HH	A2-7	A3-5034
		WE1534HH	A2-7	A3-5034
3886	1½"	WS0511A	A2-5	NR
		WE0512A	A2-6	NR
		WS0532A	A2-7	A3-5034
		WS0534A	A2-7	A3-5034
3887 ②	2"	WS0311B,BF	A2-5	NR
		WS0312B,BF	A2-6	NR
		WS0511B,BF	A2-5	NR
		WS0512B,BF	A2-6	NR
		WS0532B,BF	A2-7	A3-5034
		WS0534B,BF	A2-7	A3-5034
		WS0712B,BF	A2-7	A3-2012
		WS0732B,BF	A2-7	A3-5034
		WS0734B,BF	A2-7	A3-5034
		WS1012B,BF	A2-7	A3-2012
		WS1032B,BF	A2-7	A3-5034
		WS1034B,BF	A2-7	A3-5034
		WS1012BH,BHF	A2-7	A3-2012
		WS1032BH,BHF	A2-7	A3-5034
		WS1034BH,BHF	A2-7	A3-5034
3888DX ③	3"	WS1512D,DX	A2-7	A3-2012
		WS1532D,DX	A2-7	A3-5034
		WS1534D,DX	A2-7	A3-5034
		WS2012D,DX	A2-7	A3-2012
		WS2032D,DX	A2-7	A3-5034
		WS2034D,DX	A2-7	A3-5034
		WS3032D,DX	A2-7	A3-5034
		WS3034D,DX	A2-7	A3-5034
		WS5032D,DX	A2-7	A3-5034
		WS5034D,DX	A2-7	A3-5034

① Diaphragm Control Switch or (2) A2-3 Float Switches can be used. See Accessory page in Sewage Section if desired.

② Requires 24" Receiver Basin or larger.

③ Requires 36" Min. Dia. Receiver Basin.

NR — Not Required.

Check Valve

No.	Pipe Size
A9-2P	2"
A9-2I	
A9-2B	
A9-3P	3"
A9-3I	

P = Plastic/Flapper Style
I = Iron/Flapper Style
B = Iron/Ball Style

Sewage Basins

Model	Dia.	Depth	Capacity
A7-1801S	18"	30"	33 gal.
A7-2401S	24"	36"	72 gal.

Basin Covers

Model	Basin Dia.	Vent & Discharge Size
A8-1822	18"	2"
A8-1833	18"	3"
A8-2422S	24"	2"
A8-2433S	24"	3"

Additional Basin and Cover Sizes Available — See Catalog for Listing.

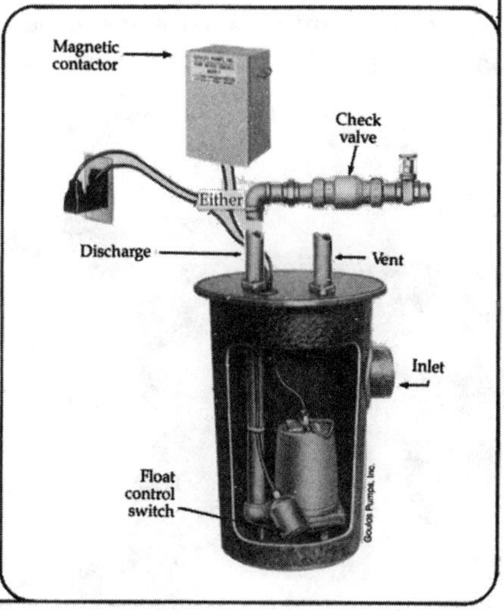

Figure 16.3 Sewer pump installation. (*Goulds Pumps, Inc.*)

There will be a 2-inch hole on the base of the pump with female threads. A male adapter is screwed into the treads to convert to glue joints for plastic pipe. The pipe from the pump will rise vertically. The electrical cord will be pulled through the top, and the top will be set down over the discharge pipe.

The top will have a waterproof gasket that it sits on to seal the sump. Screws or machine bolts will hold the top in place. The other 2-inch hole in the top will contain female threads. This is the vent hole. A male adapter is used to convert to glue joints. The vent must extend up to the open atmosphere. This means getting it from the basement to the attic and out the roof. Some areas allow these vents to be tied back into other vents; we will talk more about vents in the next chapter.

The discharge pipe will be fitted with a 2-inch gate valve and check valve (Fig. 16.4). Then the discharge pipe will enter the building drain through an approved fitting, usually either a wye or a combination wye and eighth bend.

When the sump begins to fill with waste, the float on the pump will rise, cutting the pump on. The pump will empty the contents of the sump into the building drain. The gastight gasket on the sump and the vent eliminate unpleasant odors and dangerous gas buildups.

What Is a Combination Waste and Vent System?

A combination waste and vent system is a type of plumbing system that uses very few plumbing vents. Most codes require every plumbing fixture to be vented with some type of individual vent. However, some parts of the country alter these rules significantly. Most of my plumb-

Figure 16.4 Check valve.

ing career has been spent in Virginia. It was common practice there, as it is in most states, to provide every drainage fixture with a vent. Maine, however, has a very different way of installing plumbing in most houses.

Maine has the most liberal combination waste and vent rules I have ever experienced. Essentially, this type of system relies on an oversized drain to provide both the drainage and the venting for a fixture. One 3-inch vent is required to control gas buildups and to allow some air into the system, but many fixtures can be installed without individual vents.

This combination system makes the cost of plumbing a house considerably less. You save time and money by not installing many vents. Do these systems work? Yes, but generally not as well as a vented system. Drains drain more slowly on a combination system. We will talk much more about vents in the next chapter.

Almost all codes allow for some form of combination waste and vent system, but most jurisdictions are not as liberal on the plumbing code as Maine is. There are times, such as in the case of an island fixture, where a combination system must be used. This type of arrangement gets complicated, but you will learn more about it in the next chapter.

For now, all you really need to understand about combination systems is that they exist, and they usually require larger drains than a standard system. Most of the big differences pertain to vents, and that will be better discussed in the next chapter.

Tips and Tidbits

Let's spend the remainder of our time in this chapter discussing tips and tidbits that may make your life easier. Every trade has its tricks, and plumbing is no exception. Experience is a great teacher of how to cope with problems and how to solve them more easily than you could without the experience. Let me show you what I mean.

What is dry fitting?

Dry fitting is putting pipe and fittings together without making a permanent joint for the purpose of checking the feasibility of an installation. Many plumbers and apprentices dry fit their installations before making an irreversible joint.

I have never practiced dry fitting, and I don't believe in it. In my opinion, it is a waste of time. It's not so bad for a trainee, but an experienced plumber should not have to waste time with dry fitting. In my observations of plumbers over the years, I have seen dozens of them waste enough time dry fitting every joint to have plumbed twice as many jobs. There could be rare occasions when a seasoned plumber should dry fit, but they are few and far between.

Dry fitting is a bad habit to get into. If you begin your career testing each joint before you make it, work to get out of the habit as soon as possible. Production and quality are the name of the game in competitive plumbing. If you are spending time trying to get a fitting off the end of a pipe that got stuck during dry fitting, you will be left in the dust.

Don't just stick it in; twist it

When you are gluing plastic drainage pipe into a fitting, don't just stick it in; twist it. By rotating the pipe in the fitting, you will be assuring better coverage of the glue. This will help reduce leaks.

Should you paint the hubs with glue?

When you are installing plastic drainage systems, the decision of whether or not to paint the hubs with glue is a sticky question—no pun intended. Plumbing inspectors will tell you the joint should be made with the bonding of the pipe to the inside of a fitting's hub. Many plumbers will tell you to run a bead of glue around the outside of every fitting's hub to ensure no leaks. What should you do? I agree with the inspector's point of view, but this is not to say I have never painted a hub with glue.

You shouldn't depend on a superficial seal at the outside of a hub to seal a pipe joint. If you are working with ABS pipe, you will never have to worry about it one way or the other; it is almost impossible to make ABS leak if you get glue on the pipe and fitting. However, PVC pipe does tend to leak more easily. The leaks are usually the result of the pipe and fitting not being property prepared with cleaner and primer; why do you think the code wants you to use the stuff? If you make a PVC joint in the proper manner, it shouldn't leak.

I'm sure you will be told at some point during your plumbing career to paint your hubs with glue. The choice is yours, but I think you will be better off to do the job the right way.

Use the right adapters

If you have to mix different types of pipe, use the right adapters. You might use threaded adapters, rubber couplings (Fig. 16.5), or other special adapters but be sure they are the adapters intended for the use you are putting them to. The wrong adapters can result in leaks and failed inspections.

Figure 16.5 Rubber coupling.

Adequate space

When you are roughing-in drains, you must make sure there is adequate space for the fixture you are roughing-in. Not all builders and carpenters are aware of the space requirements for specific plumbing fixtures. You are the plumber; it is up to you to tell the general contractor if the space provided for your fixtures is inadequate. These space requirements vary to some extent with different codes, but I can give you some samples of what to look for. The requirements are different for handicap fixtures, which we will cover after the standard fixtures.

Space requirements for standard fixtures. The space requirements for standard fixtures are not hard to identify or understand, but they are important to obey. Let me tell you what to look for.

Fixtures, such as urinals, bidets, and toilets, should have a minimum open distance in front of them of at least 18 inches (Fig. 16.6). This measurement is made from the front edge of the fixture to the first object in front of it.

You should be able to measure 15 inches to either side of the center of a drain for a toilet, bidet, or urinal without hitting another object. When these fixtures are installed side by side, there should be a minimum of 30 inches between the centers of the two drains.

Privacy stalls built around toilets must have a width of at least 30 inches (Fig. 16.7) and a depth of at least 5 feet.

There should be at least 18 inches from the apron of a bathtub to another object, whether it is another fixture or a wall.

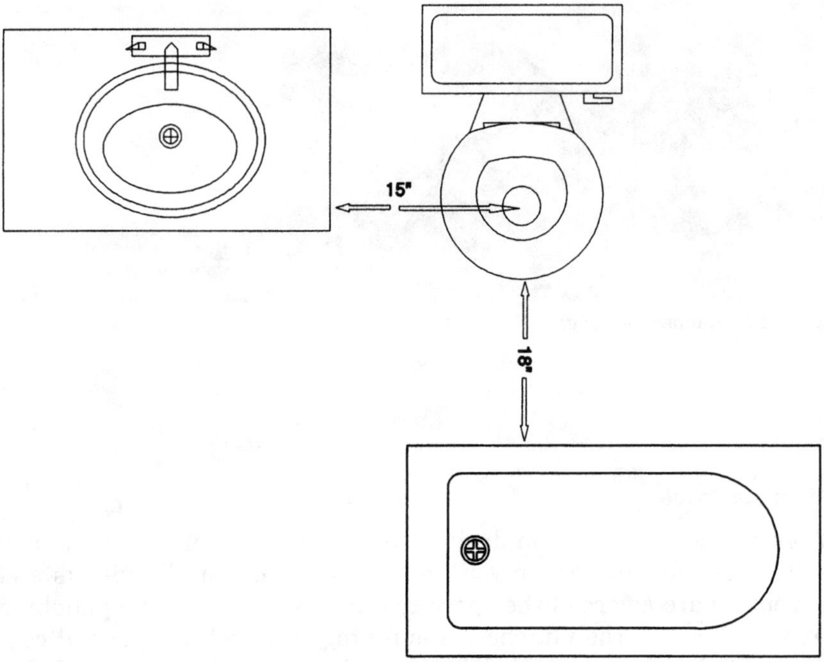

Figure 16.6 Minimum distances for legal layout.

Handicap fixture spacing. Handicap fixture spacing is a bit different. Like the other spacing requirements, these requirements vary from code to code. I will not attempt to cover all of the measurements and possibilities, but I will give you some solid examples.

Water closet compartments for handicap toilets must be at least 5 feet wide. The door is required to swing out, away from the toilet, and it must present a clear opening of at least 32 inches. There must be a clear space of at least 48 inches in front of the compartment door. Grab bars are required in these enclosures. There are many different shapes and dimensions that can be used to make toilet compartments comply with handicap regulations. Check your local code for specifics.

The clear space required in front of a handicap lavatory is 30 inches. There are also regulations that deal with the height of the lavatory top and toe clearance. Some more information will be found on this subject in Chapter 20, but you must check with your local code requirements for all pertinent details.

The minimum clear space in front of a handicap bathtub must be at least 30 inches. This clear space must have a width of at least 4 feet.

As you can see, size requirements for handicap fixtures are significantly different from those of standard fixtures. To be safe, refer to your

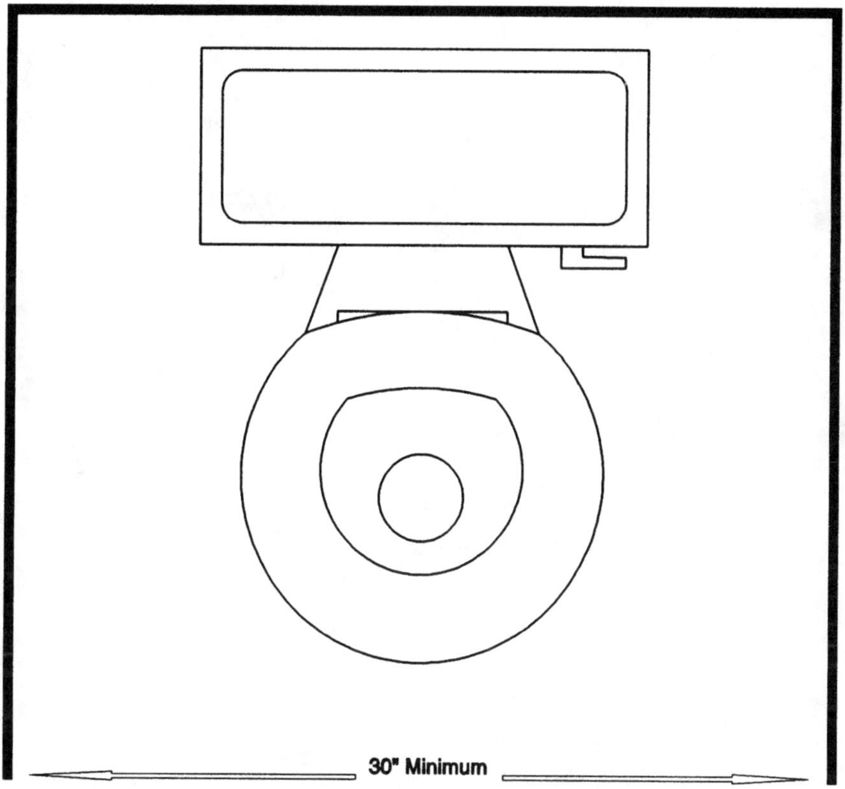

Figure 16.7 Minimum width requirements for a water closet.

local plumbing and building codes for precise instructions on the placement of your rough-ins.

Rough-in measurements

Rough-in measurements are a common concern for inexperienced plumbers. Never fear; there are entire books that do nothing but show you the proper positions for roughing-in fixtures. Some rule-of-thumb figures for rough-ins are included in the chapter on fixtures: Chapter 20.

Follow the code

To keep yourself out of trouble, always follow the code. The plumbing code was drafted for good reasons, and it is your duty and responsibility as a plumber to adhere to it. Now, let's move on to venting and see what's in store for us there.

Chapter

17

Vent Systems

Vent systems are very important to plumbing systems. However, many people, some plumbers included, don't understand why vents are needed or how many jobs they do. Most everyone assumes that plumbing vents are used to keep foul odors from invading habitable space. This is one reason for venting, but it is not nearly the only one. Vents protect trap seals from siphonage, remove dangerous gases from the plumbing pipes, allow drains to flow better, and control odors.

There are many types of vents. There are stack vents, vent stacks, individual vents, common vents, wet vents, circuit vents, and others. Each of these vents has its own special place in plumbing.

Since most jurisdictions require every plumbing fixture to be vented, venting is a big part of the plumbing trade. It is a part of the job that confuses many people. The reasons for installing vents are easy to identify, but it is not always so easy to decide on what type of vent to use. Even after a decision is made on which style of vent is best suited to the job, some plumbers have trouble understanding the code requirements for installing the vents.

There is no big mystery to proper venting procedures. The plumbing code books dedicate adequate space to the subject. There have been many times that I've thought plumbers have trouble grasping venting procedures because they don't think vents are important enough to concern themselves with. This thought bothers me, because I know the importance of vents.

This chapter is going to explain a lot to you about vents and venting. Apprentices are not expected early in their career to know all there is to know about vents. While it is not your responsibility to understand all aspects of venting, you should know the principles and basics.

Before you can get your plumber's license, you will have to understand venting, so you might as well get a head start on it.

I gave you on-the-job examples near the beginnings of the last two chapters. The physical installation of vents is very similar to piping a drainage system. The primary difference is that drains go down and vents go up. Since the general work is so closely related to drainage work, we will skip the job-site example in this chapter. There is much to learn about vents, so let's get started.

What Is a Plumbing Vent?

A plumbing vent is a device, normally a pipe, that allows air into a plumbing system and removes gases from the system. Most vent pipes terminate into the open atmosphere at a point above the roof of a structure.

Why Are Vents Needed?

Vents serve multiple functions. A plumbing vent will carry potentially dangerous gases into the open air where their danger is not a threat. Fixtures with nearby vents drain much faster than fixtures without vents. The vents allow air circulation that increases the flow of drainage. It is similar to trying to pour liquid from a can that has only one hole in its top. If you punch a second hole in the top, the liquid flows more smoothly and faster. The vent is often thought of as preventing odors from entering a building, but in reality, trap seals do most of that job. Vents, however, protect trap seals, so in a way, they do work to keep odors out of a building.

How Do Vents Protect Trap Seals?

Vents allow air to flow in the plumbing system, reducing the risk of siphonage. If siphonage is eliminated, trap seals are safe, as long as the fixture is used periodically or is equipped with a trap primer.

Most traps can lose their water seals if a pressure change in the plumbing system occurs. The force of the pressure can suck the water out of the trap and down the drain. Vents act to prevent extreme pressure changes and back siphonage, thereby protecting the trap seals.

If a fixture is not used from time to time, the water in the trap can evaporate. If the water in the trap is lost to evaporation, sewer gas and foul odors can escape through the fixture into the living space. This creates health hazards.

Fixtures that see little use, like some floor drains, are equipped with trap-seal primers. These devices fill the traps with water periodically, eliminating the risk of total evaporation of the trap seal.

What Fixtures Should Be Vented?

All fixtures should be vented. However, with some plumbing codes, a fixture's drain acts as its vent. This type of arrangement is known as a combination waste and vent system. Most codes have some allowance for venting through the combination system, but some are much more liberal than others. Ideally, every fixture should be vented with a vent going to the outside air. The vents should be installed as close to the fixtures as possible.

Isn't Venting Every Fixture a Waste of Pipe?

Some plumbers believe it is a waste of time and material to vent every fixture. Other plumbers, like myself, see the benefit of venting all fixtures and believe every fixture should have its own vent. However, there is room for argument on both sides of the debate.

Plumbers who have worked with combination waste and vent systems for years will swear that they work well and that there is no need for a bunch of vents running into the attic and out of the roof. But when a plumber who is used to venting every fixture is asked the same question, you will get a different answer. Plumbers who are accustomed to vented systems can't believe how slowly bathtubs drain when they are a part of a combination waste and vent system.

As an example, I am used to seeing the tiny tornados made by water draining out of vented bathtubs. When I work with a tub on a combination system and watch the water drain, it is like watching a mud puddle evaporate; it's a slow process.

Plumbers who don't believe in venting every fixture will argue that the requirements for so many vents is a result of suppliers wanting to sell more pipe. When requirements call for full venting, plumbers must spend more time on the job. This makes the jobs more expensive, but it also keeps the plumbers working and making more money.

Whether vents are required out of greed, politics, or need, there is no doubt that they are beneficial to a plumbing system. If you don't believe they are, visit houses that are plumbed with and without vents. Then decide which house you would prefer to live in.

And, by the way, venting every fixture doesn't mean running individual pipes from every fixture to the roof of the building. This is a common misconception of plumbers who are not used to full-scale venting. There are many ways to use wet vents, circuit vents, and other vents to reduce the amount of pipe needed in a fully vented system.

What Are the Different Types of Vents?

The types of vents available to a plumber are numerous. Each type usually has its own specialized purpose. Let's look at each of the most common types of vents and see when they are used.

Branch vent

Branch vents (Fig. 17.1) are used to connect any number of individual vents to vent stacks or stack vents. This is one example of how the volume of pipe used to vent a system can be reduced. Branch vents can be used to run a vent horizontally to another vent.

How does this save pipe? It's simple; consider this example. You are plumbing a bathroom in the basement of a home. There is a 3-inch stack vent that goes up the wall behind the toilet. The stack vent is

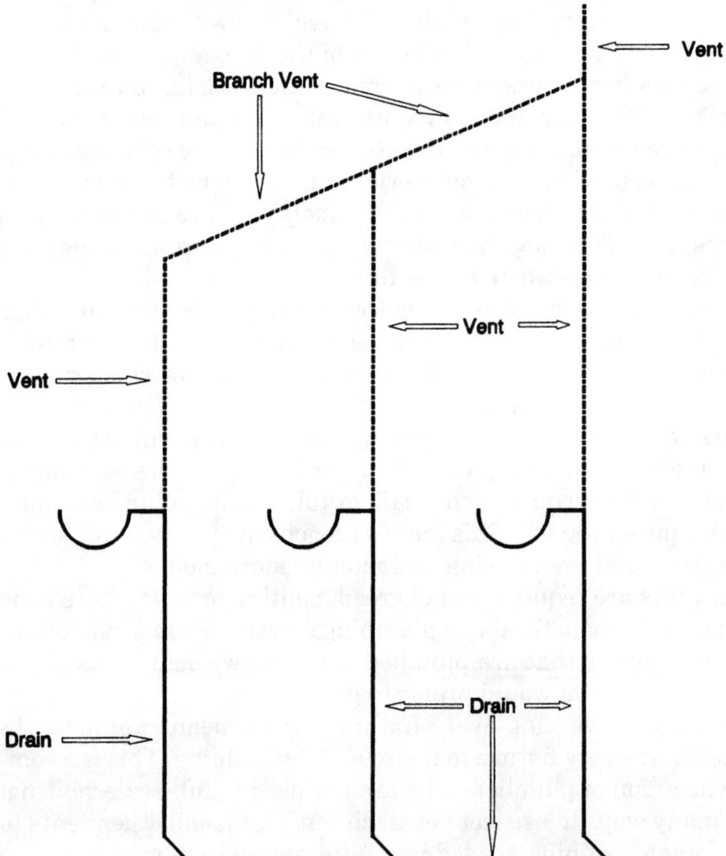

Figure 17.1 Branch vent.

going to extend up through the house and exit the roof. Your bathtub is wet vented through the lavatory. (I know you don't understand all of these terms yet, but you don't need to for this example. I will explain the terms a little later.) You have a 2-inch pipe that serves as your lavatory drain and the wet vent of the bathtub. That 2-inch drain must have a vent that comes from the top of it and releases its contents into the open atmosphere.

You could extend the vent up from the 2-inch pipe until it penetrated the roof. If you do this, you have two vents going from the basement bath to the roof. This would be a waste of pipe. However, you can use a branch vent to reduce the waste. After raising the lavatory vent above the flood-level rim of the fixture, usually by at least 6 inches, you can make a 90-degree turn with the vent and tie it into the 3-inch pipe. The 3-inch stack vent will assume the responsibility for transferring the contents of the lavatory vent to an open air space.

The vertical part of your lavatory vent is an individual vent. The horizontal run of pipe is the branch vent. Installing the venting in this manner should save at least 20 feet of pipe and probably more. It will also save you time and effort.

Some plumbing codes allow horizontal vents to be installed level, and other codes require that horizontal vents be graded. Even when the code doesn't require vents to be graded, they should be. A horizontal vent that is installed level can retain rain water, leaves, and other debris the might enter the system. This could cause the vent to become blocked and ineffective. If the vent is pitched upward, any water or debris will flow down the vent, toward the drainage system. Most good plumbers install their vents with a upward grade of 1/4 inch to the foot.

Circuit vent

A circuit vent (Fig. 17.2) is a type of branch vent. It also serves to reduce the amount of pipe required for full-scale venting. These vents serve multiple fixtures and extend from just in front of the last fixture being served to return to the vent stack. In other words, circuit vents allow a number of fixtures to be vented with a minimum of pipe.

To visualize a circuit vent, think about a job where there are five floor drains in a row, spaced about 10 feet apart. Each trap is considered a fixture and will normally require a vent. You can run five individual vents to the roof, or you can use a circuit vent.

In this installation there is a vent stack at the drainage end of the floor drains. When the drains are installed, a vent rises up from the drain, just in front of the last floor drain in the line. The vent rises and then turns on a 90-degree angle, headed back to the vent stack. The vent ties into the vertical vent stack and all of the floor drains are vented with the use of the circuit vent.

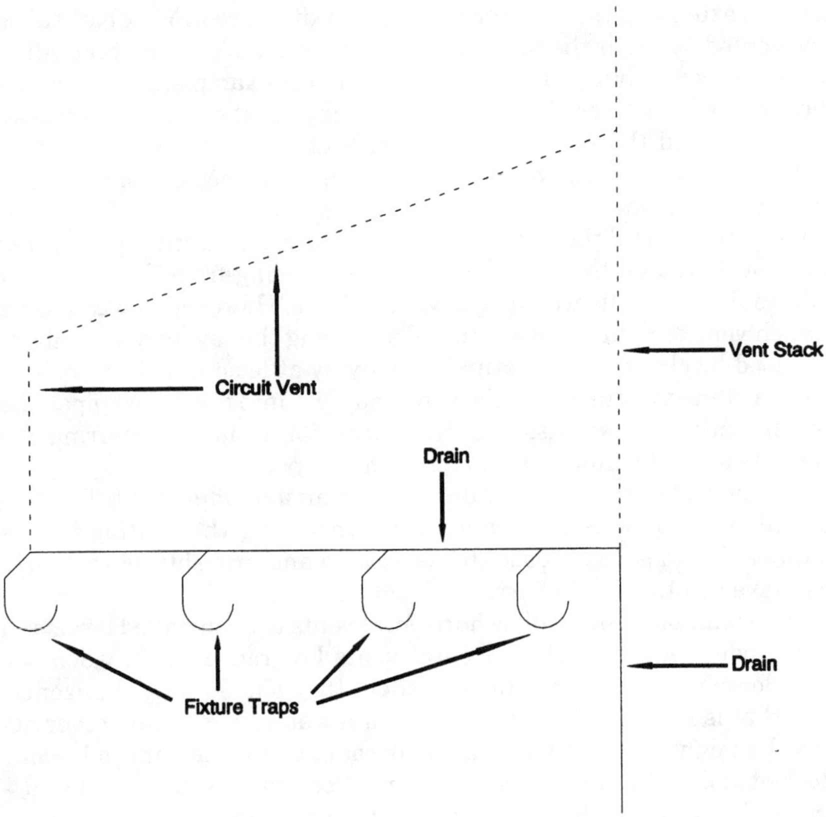

Figure 17.2 Circuit vent.

Common vent

A common vent (Fig. 17.3) is one vent that is serving multiple traps. A common vent could be used to vent two lavatories that were installed, back-to-back, in separate bathrooms.

Continuous vent

A continuous vent is simply a vertical vent that is an extension of the drain it is serving. These vents rise vertically until they exit into open air. While we are talking about vertical vents, let me tell you about 45-degree offsets. If a pipe is rising vertically and is offset with the use of two 45-degree fittings, resulting in a vertical climb, it can still be considered a vertical pipe.

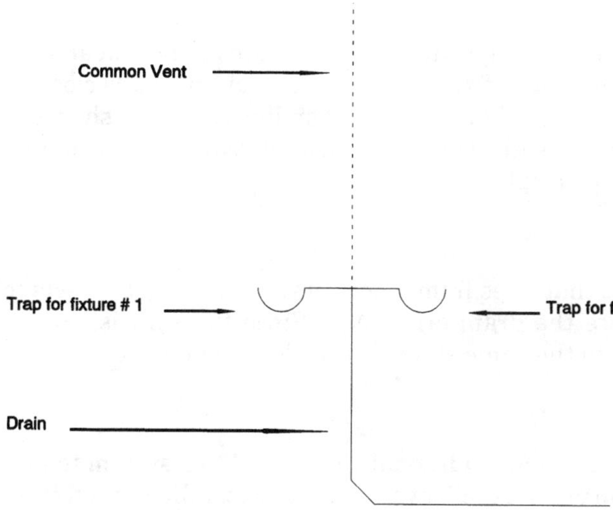

Figure 17.3 Common vent.

Individual vent

Individual vents (Fig. 17.4) serve only one fixture. These vents can connect to other vents in the plumbing system, or they can extend to open air on their own. These vents are common on kitchen sinks.

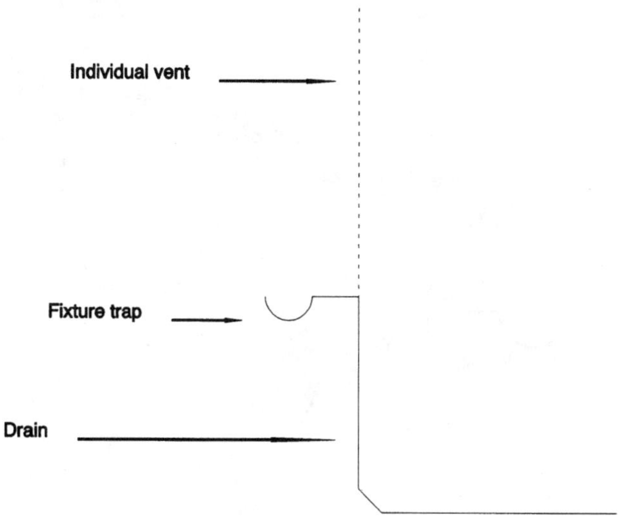

Figure 17.4 Individual vent.

Island vent

Island vents are the most complicated vents you are likely to deal with. These vents are used to vent fixtures when there are no walls close by in which to hide a vent pipe. I have included an illustration to show you what an island vent looks like (Fig. 17.5). Island vents use a combination waste and vent principle.

Loop vent

A loop vent is a vent that rises from a horizontal drain and connects to the stack vent where the drain originated. Since the vent is looping back over the drain to the same stack, it is called a loop vent.

Relief vent

A relief vent (Fig. 17.6) is an additional vent added to a system to provide more circulation of air. A relief vent might be installed on a fixture branch if the branch was going to exceed the allowable distance between the trap and the vertical vent.

Relief vents are also used on drainage stacks that have more than 10

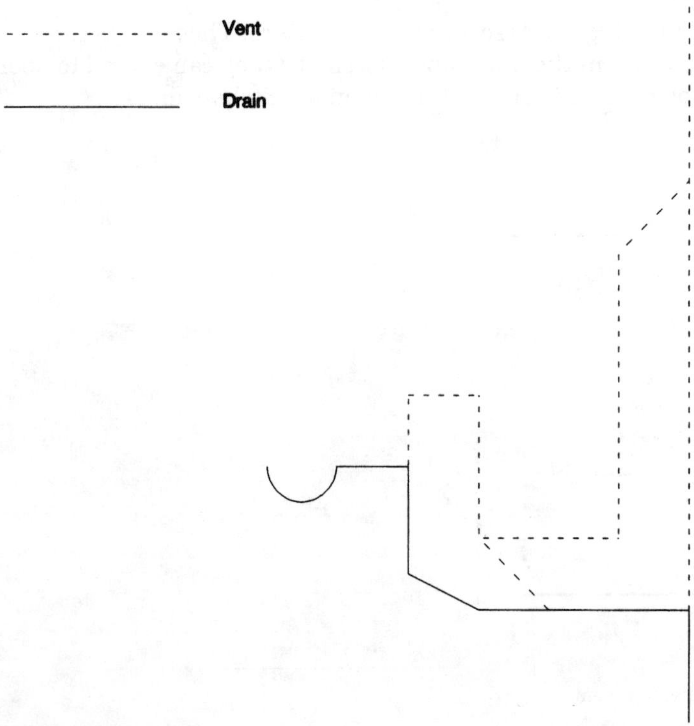

Figure 17.5 Island vent.

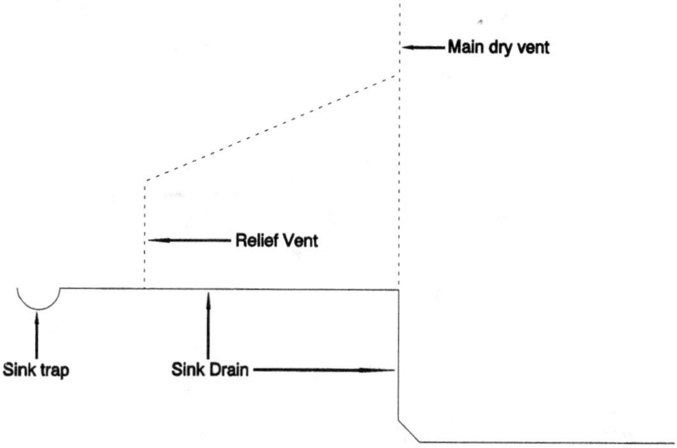

Figure 17.6 Relief vent.

branch intervals. The vents are installed at every tenth interval and tie into a vent stack.

Stack vent

A stack vent (Fig. 17.7) is made when a waste or soil stack rises above the highest drainage fixture and seeks open air. The part of the pipe at or below the highest drainage fitting is a drain. The part of the pipe above the drainage fitting is a vent.

Vent stack

Vent stacks (Figs. 17.8 and 17.9) are vertical pipes that carry no drainage. These pipes are run in the vicinity of soil and waste stacks and accept the contents of various vents that tie into them. Vent stacks may tie into a soil stack or originate from the building drain and extend to open air.

If a vent stack connects to a soil stack, the connection must be made at or below the lowest branch interval. When the connection is made to the building drain, it must be made within 10 pipe diameters of the drainage stack and on the downstream side of the stack. For example, a 3-inch vent stack would have to connect to the building drain within 30 inches of the soil or waste stack, and it must be on the downstream side of the stack.

Wet vent

Wet vents are another method of venting that conserves pipe. Wet vents are pipes that serve as the drain of one fixture while venting an-

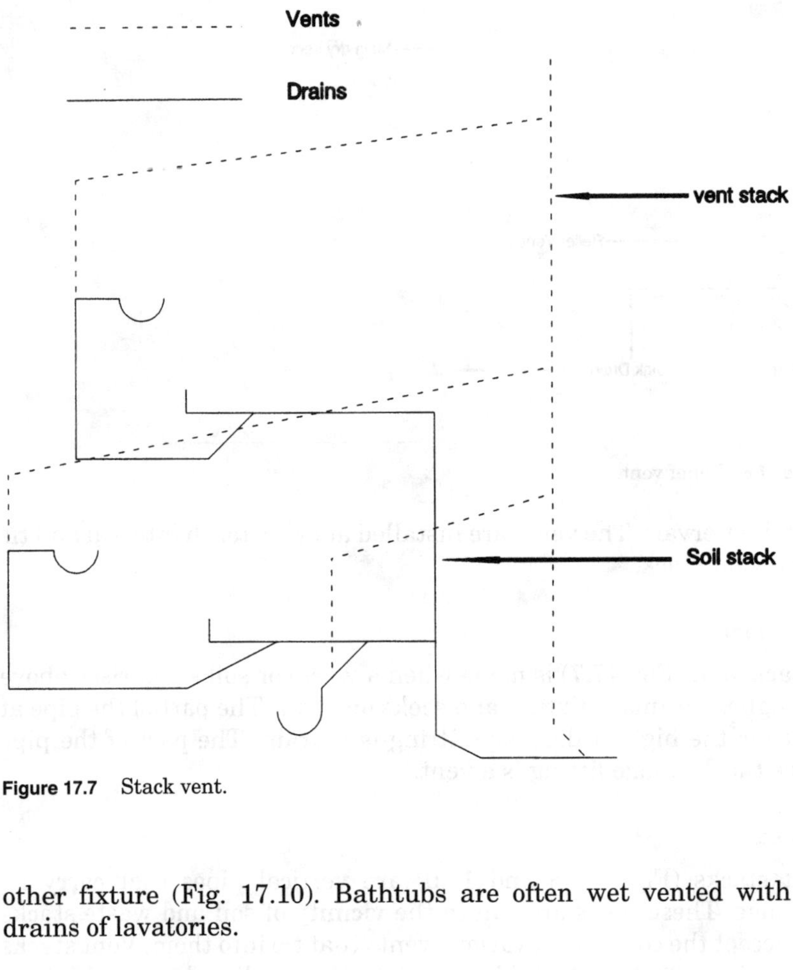

Figure 17.7 Stack vent.

other fixture (Fig. 17.10). Bathtubs are often wet vented with the drains of lavatories.

Yoke vent

Yoke vents (Fig. 17.11) are vents that connect upward from a soil or waste stack to a vent stack. These vents are installed to prevent pressure changes in the stack, but they are not used very often.

Can All Vents Tie into Other Vents?

Generally, all vents can tie into other vents (Fig. 17.12). However, there are exceptions. For example, if you are on a commercial job where there is a piping system of chemical wastes, the vents from the chemical waste system cannot tie into the vents of the sanitary drainage system.

Unless you are working with special wastes, such as you might find in a hospital, medical offices, or factories, you can feel pretty safe in

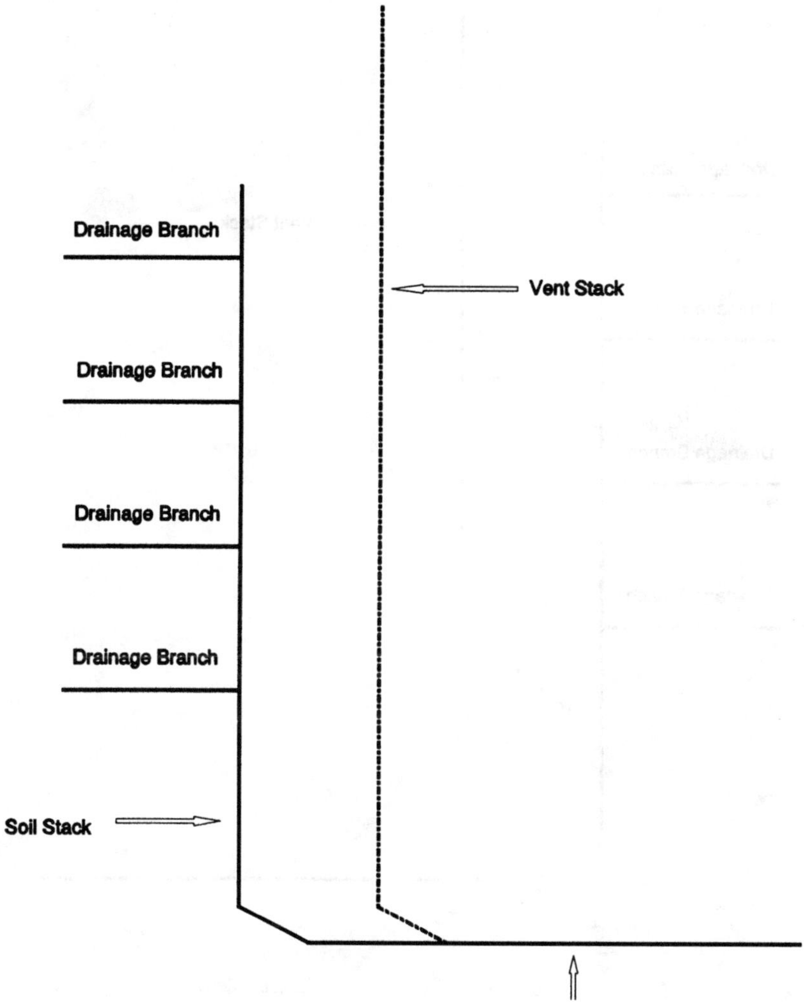

Figure 17.8 Vent stack.

tying any vent back to the main vent of a plumbing system. If you are working with special wastes, consult your code book, master plumber, or plumbing inspector.

When Would I Use a Vent Stack?

Vent stacks can be used in any standard plumbing system. They are usually required if the building has a drainage stack accepting five or more branch intervals. As you may remember, branch intervals are the horizontal drains extending directly from a stack.

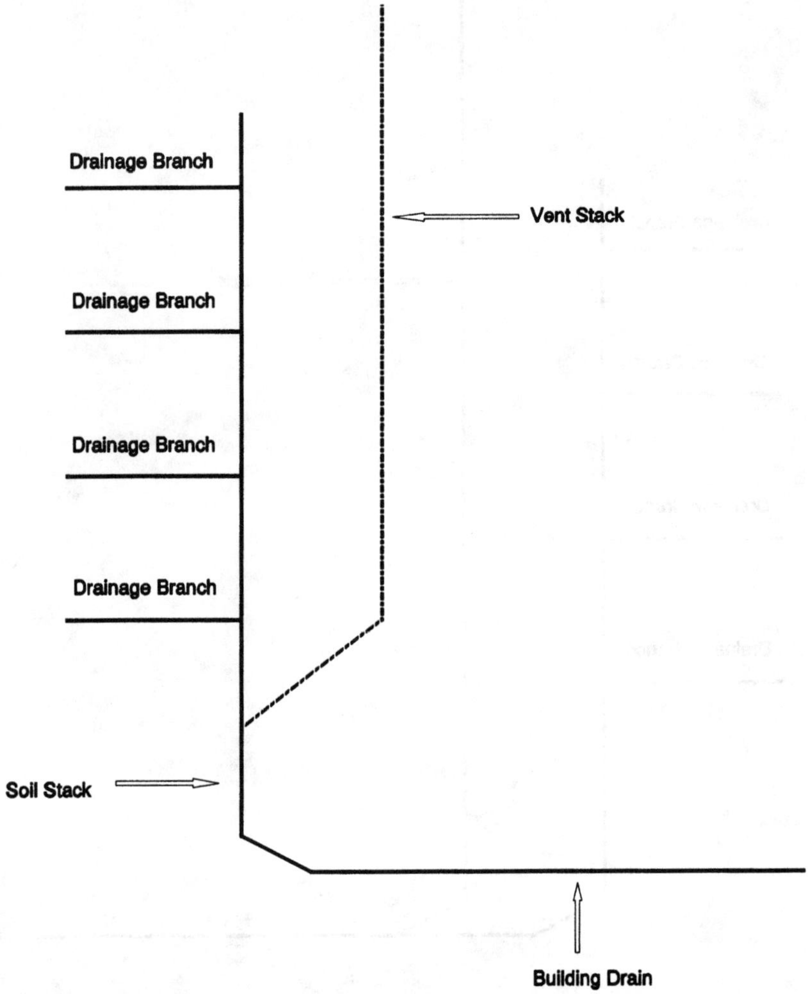

Figure 17.9 Vent stack.

Can Vents Be Stopped Once They Are in an Attic?

Stack vents and vent stacks must penetrate into the open atmosphere, outside of the building. Other vents can tie into these vents, but no vent may be left open in an attic or any other part of a building.

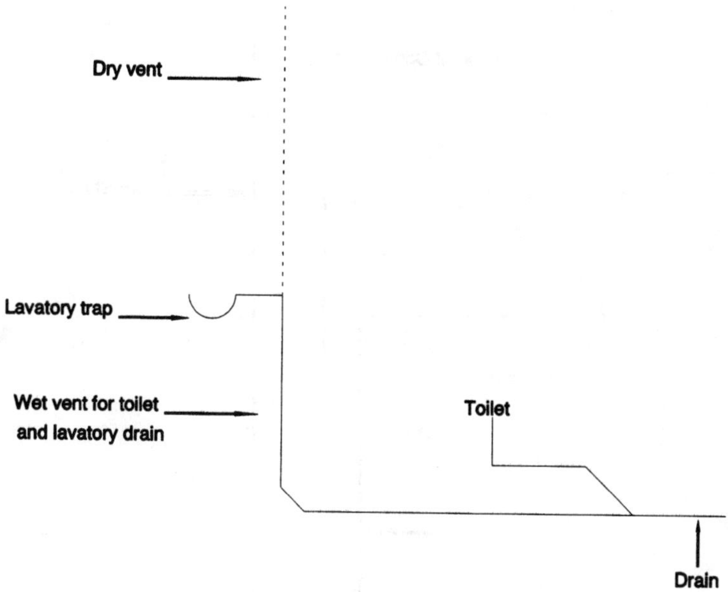

Figure 17.10 Wet vent.

Minimum Requirements

The minimum requirements for a plumbing system that collects the discharge from a toilet call for at least one 3-inch vent stack or stack vent. This main vent must connect to a drainage pipe with a diameter of at least 3 inches, usually the building drain.

An Example of When a Yoke Vent Might Be Used

If you were working with a drainage stack that was offset into a horizontal position, you might use a yoke vent. If the stack had five or more branch intervals above where the offset took place, you would have to install a vent stack and use a yoke vent to vent the stack offset. You could run into this situation on big jobs, but the occasions will be rare.

Where Do Vent Pipes Terminate?

Vent pipes either tie into main vents or terminate into the open air outside a building. The length of extension past the roof of the building

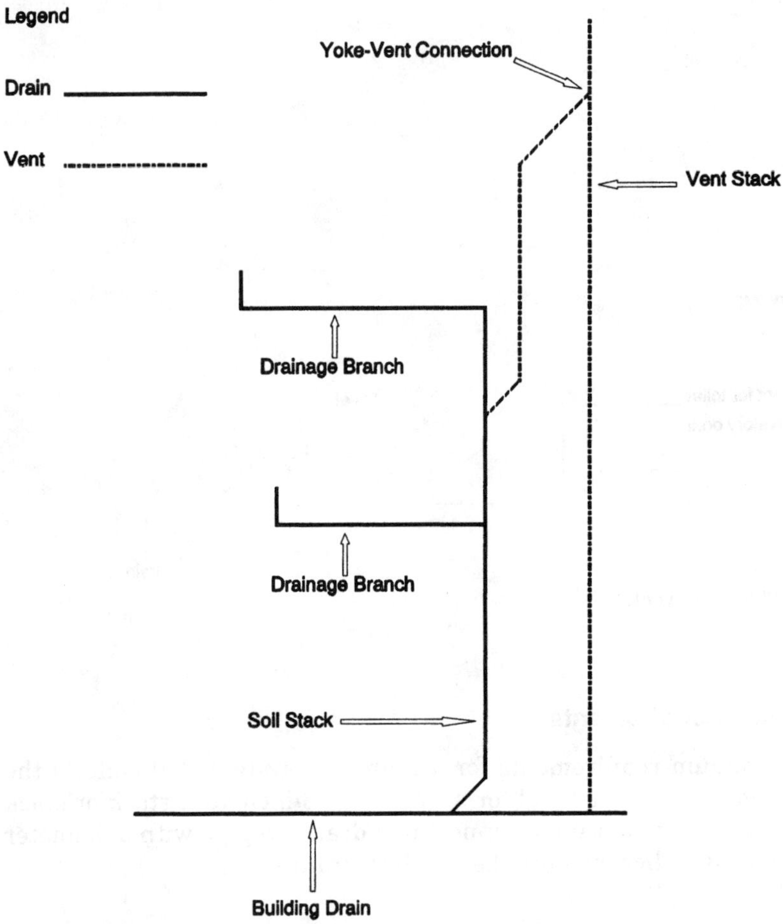

Figure 17.11 Yoke vent.

varies. In most cases vents stop about 1 foot above the roof. However, some codes require the vents to rise 2 feet above the roof. Other codes require only a 6-inch extension above the roof. This is just another example of why you must always refer to your local code requirements before installing plumbing.

If a roof is being used for something other than weather protection, such as a roof-top patio, the vent must extend at least 7 feet above the roof. It is also important to keep vents away from windows and other openings in the building. Vents are generally required to be at least 10 feet away, measured horizontally, from any such opening. If it cannot be kept away from the opening, the vent must extend at least 2 feet above the window, door, or other opening (Fig. 17.13).

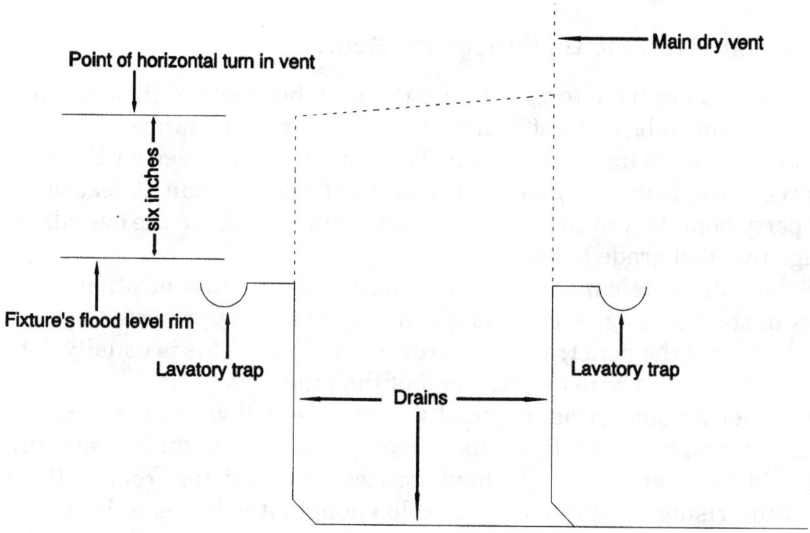

Figure 17.12 Vents tying into each other.

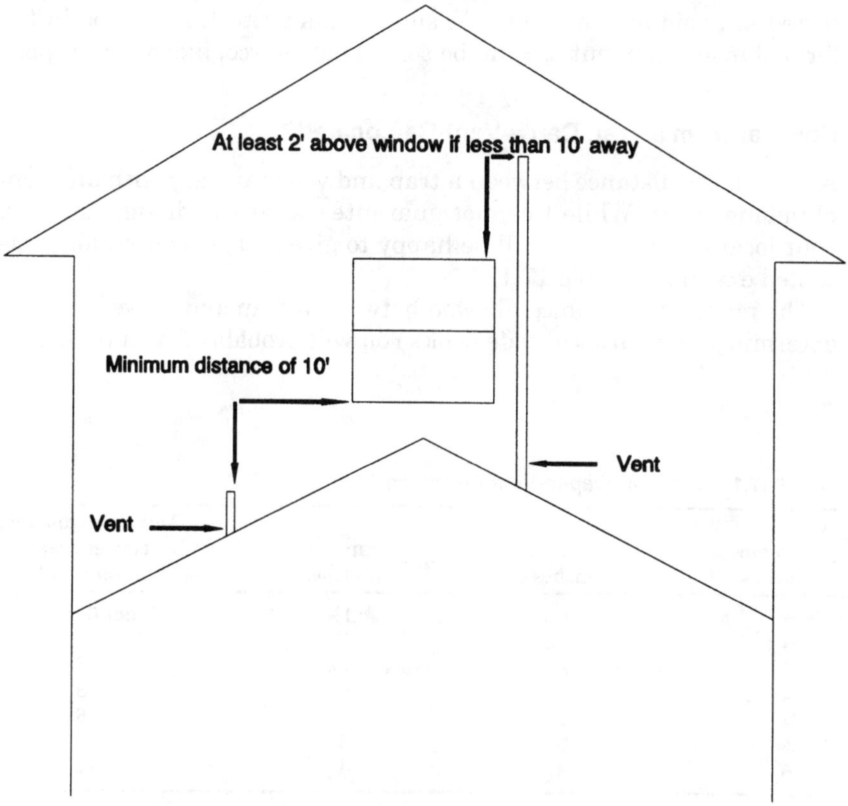

Figure 17.13 Venting rules for termination.

Do All Vents Have to Go through the Roof?

All vents do not have to go through the roof, but most of them do. It is usually allowable, at least under extenuating circumstances, to take a vent out through an exterior wall. There are varying rules for this type of procedure. Normally, the vent must not be closer than 10 feet to the property boundary of the property, and it must be above the overall average finished grade level.

Side-wall vents cannot terminate under soffit vents or other openings in the building. Additionally, the pipe must be protected against blockage and the penetration of birds and rodents. This is usually done by placing screen wire over the end of the pipe.

Another consideration, especially with side-wall vents, is the risk of freezing temperatures. If you have ever looked at a plumbing vent during cold weather, you might have noticed vapors rising from it. If the moisture rising in vent pipes gets cold enough, it will freeze. When the temperatures stay below freezing for extended periods, the freezing moisture can accumulate and block the opening of the pipe, rendering the vent useless. If there is a risk of freezing, the vent pipe must be protected to avoid an ice buildup. Usually, this protection will be in the form of insulation, but it could be some heat source, like a heat tape.

How Far from a Trap Can a Vent Be Located?

As usual, the distance between a trap and vent can vary with different plumbing codes. While I cannot guarantee what the distances are in your local jurisdiction, I will be happy to give you some commonly accepted examples (Table 17.1).

The maximum distance allowed between a trap and its vent will be determined in your local code book. You will probably find a table that

TABLE 17.1 Common Trap-to-Vent Distances

Grade on drain pipe, in inches	Fixture's drain size, inches	Trap size, in inches	Maximum distance between trap and vent, in feet
¼	1¼	1¼	**3 feet 6 inches**
¼	1½	1¼	5
¼	1½	1½	5
¼	2	1½	8
¼	2	2	6
⅛	3	3	10
⅛	4	4	12

shows you what is allowed. If the information is not in a table, it will be in the text.

Let's say you have an 1½-inch trap and an 1½-inch drain; the vent can often be up to 42 inches away from the trap. If you were working with a 2-inch trap and drain, the allowable distance might be 6 feet. Similar circumstances involving 3-inch pipe and drains could allow for a distance of 10 feet between the trap and the vent. A 4-inch trap and drain may allow a distance of up to 12 feet.

When you are computing these distances, you must measure the total developed length between the drain from the vent to the trap. This would include the length of the trap arm—the piece of pipe that normally sticks out of the wall and connects the trap to the drain.

Can the Vent Rise Right from The Trap?

As a rule, the vent cannot rise right from the trap. This practice is known as crown venting. Crown-vented traps are normally prohibited. However, it may be acceptable to extend the vent from the trap arm. The vent should be at a distance from the trap weir equal to at least two pipe sizes (Fig. 17.14). In other words, a trap-arm crown vent for a 2-inch trap would have to be at least 4 inches away from the weir of the trap.

How Do You Know What Size Vent Pipe to Use?

You will use the text and tables in your local code book to size your vent pipes (Tables 17.2 and 17.3). The sizing requirements vary from code to code and for different types of vents. Let me show you how some sizing examples work out.

Sizing vent stacks and stack vents

When sizing vent stacks and stack vents, you must know the total number of fixture units that will be connected to the stack. You will also have to know the size of the drainage pipe being used and how long the vent pipe will be. Once you have gathered this information, sizing the vent is easy.

You will refer to your code book for specific requirements, but this is how one sizing method might work out. Let's say you are dealing with a drain pipe with a 2-inch diameter. The total number of fixture units that will be connected to the vent is 11. If you run a 1¼-inch vent pipe, you can extend the vent 30 feet. A 1½-inch pipe will allow the vent to run 75 feet—quite a difference. A 2-inch pipe used for the vent extends the approved distance to 200 feet.

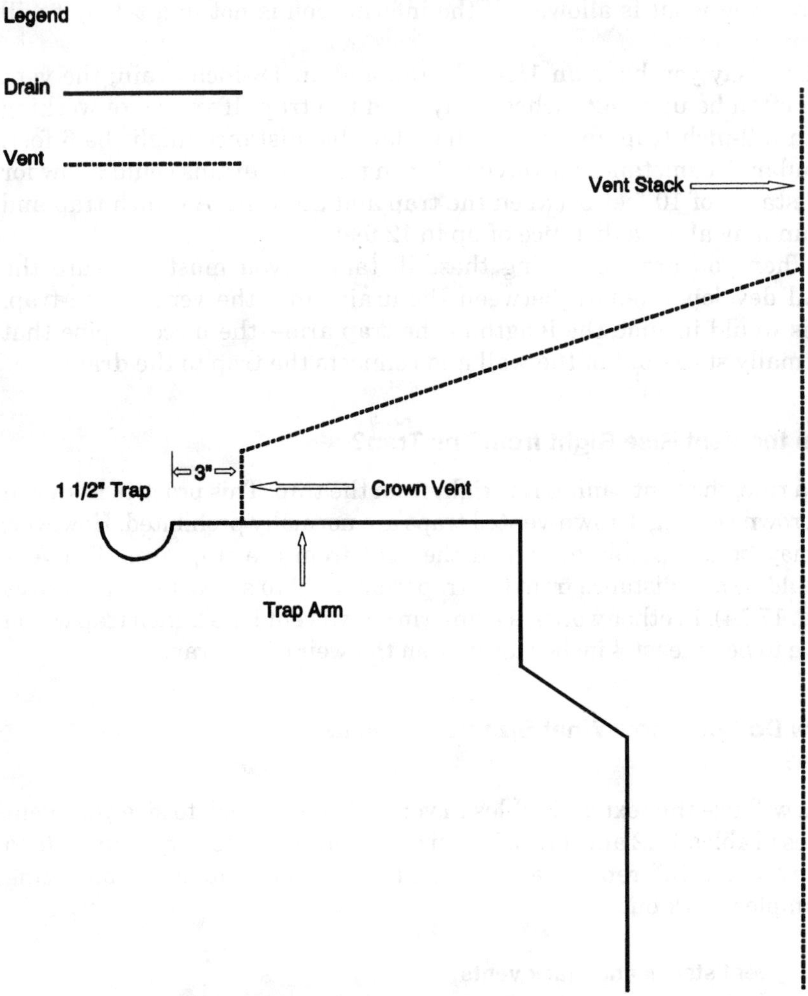

Figure 17.14 Crown venting on a trap arm.

Let's look at the same type of setup with a 3-inch drain and a total of twenty-one fixture units. A 1¼-inch vent is too small and cannot be used. Vents are normally no smaller than one-half the size of the drain. A 1½-inch vent pipe allows a run of only 32 feet, but this is enough under most residential conditions. A 2-inch pipe would allow the vent to extend out to 110 feet. To obtain a running distance of 810 feet, you would have to vent the drain with a 3-inch pipe.

As you can see, when the vent is the same size as the drain, it can run

TABLE 17.2 Common Vent Sizing Table*

Drain pipe size, in inches	Fixture-unit load on drain pipe	Vent pipe size, in inches	Maximum developed length of vent pipe, in feet
11/2	8	11/4	50
11/2	8	11/2	150
11/2	10	11/4	30
11/2	10	11/2	100
2	12	11/2	75
2	12	2	200
2	20	11/2	50
2	20	2	150
3	10	11/2	42
3	10	2	150
3	10	3	1040
3	21	11/2	32
3	21	2	110
3	21	3	810
3	102	11/2	25
3	102	2	86
3	102	3	620
4	43	2	35
4	43	3	250
4	43	4	980
4	540	2	21
4	540	3	150
4	540	4	580

*For use with vent stacks and stack vents.

TABLE 17.3 Typical Vent Sizing Table*

Drain pipe size, in inches	Drain pipe grade per foot, in inches	Vent pipe size, in inches	Maximum developed length of vent pipe, in feet
11/2	1/4	11/4	Unlimited
11/2	1/4	11/2	Unlimited
2	1/4	11/4	290
2	1/4	11/2	Unlimited
3	1/4	11/2	97
3	1/4	2	420
3	1/4	3	Unlimited
4	1/4	2	98
4	1/4	3	Unlimited
4	1/4	4	Unlimited

*For use with individual, branch, and circuit vents for horizontal drain pipes.

a long way. This type of sizing is simple once you have found the table or text in your code book.

Wet vents

Wet vents are sized based on the number of fixture units they receive. There will be a table or text in your code book to explain local options. For example, a 1½-inch wet vent can only carry one fixture unit in some jurisdictions. A 2-inch wet vent might be allowed to handle up to four fixture units. If you are looking for a wet vent that can take up to 12 fixture units, look to 3-inch pipe.

Common vents

Common vents are sized somewhat like wet vents. Once you establish the number of fixture units dumping into the piping system from the highest fixture being vented, you can cross-reference a chart or text to determine the minimum pipe size for your vent.

Sizing other vents

Sizing other vents is done in about the same way as sizing stack vents and vent stacks, but different rules apply. You will find either a separate table or other text to describe the allowable distances for vents such as circuit vents, individual vents, branch vents, and so forth.

Are Vents from Sewer Sumps Considered Regular Vents?

Usually vents from sewer sumps are considered regular vents, but it depends on the type of pump that is installed in the sump. For standard, nonpneumatic pumps, the vent from the sump can be treated much like any other sanitary drainage vent. The sizing for these vents is based on developed length and the discharge capacity of the pump.

A sewer pump's discharge capacity will be rated in gallons per minute. In residential situations, you will almost always be covered if you install a 2-inch vent. The tops for most residential sumps are made to accept 2-inch pipes, so there is not much to remember.

There are tables or text in your local code that will address the sizing of vents for sewer sumps. You will check the gallon per minute rating of the pump and compare it to the developed length the vent will run. This approach will give you the answer to your sizing questions.

Pneumatic pumps

Sewer sumps that contain pneumatic pumps require a different type of venting. The relief pipe from a pneumatic sewer pump will normally be required to connect to an independent vent stack that terminates above the roof of the building. This pipe may not be smaller than 1¼ inch. It must be large enough to relieve air pressure in the pump to a point equal to atmospheric pressure.

Who Is Responsible for Sealing the Vent Holes in the Roof?

Roofing contractors often install the pipe flashings for vent pipes, but the responsibility of the job rests on the plumber's shoulders. If the roof has already received its final covering when the holes are cut for vent pipes, the plumbers typically install the flashings. If the roof has not received it's final covering, the plumbers usually leave the flashings on the job for the roofers to install.

Most flashings for vent pipes have a metal body with a rubber collar, or they are made completely from a flexible, plastic-like material. Some roof flashings are formed from lead or copper, but these types of flashings are not too common, except on some commercial jobs.

More about Combination Waste And Vent Systems

I will tell you more about combination waste and vent systems, but be advised, the discrepancies in the rules for these systems vary widely between local codes. Combination waste and vent systems require an entirely different mind set from regular plumbing. The techniques used are drastically different. In many ways a combination system is easier to install; you don't have to extend nearly as many vents. You will, however, have to install larger drains.

There are three major plumbing codes in use today, and there are many other custom-made codes in existence with local jurisdictions. Since it would be impossible for me to know everything about every code, the information I will provide you with may not be totally accurate in your jurisdiction. However, in fairness, I will give you examples of several different concepts of combination systems.

Most of the examples given of plumbing codes up to this point have been based on the code I have worked the most with. In this section, you will get a glimpse at how all the major codes deal with combination systems. I will even give you some detailed information on how liberal Maine's alternative code is.

One version of a combination system

We will start our exploration with one version of an East Coast combination system. The types of fixtures that can be included in this type of system are lavatories, sinks, floor drains, and standpipes. In this version of a combination system, vertical pipes must be connected between fixture branches and the horizontal drainage and vent pipes. These vertical risers may not extend above 8 feet. The drainage and vent pipe may not have a grade of more than ½ inch per foot.

The riser may have its own vent, or it may connect to a horizontal drainage and vent pipe that is vented. If the riser is vented directly, the vent must rise at least 6 inches above the flood level of the fixture before turning horizontally.

Sizing for this system is determined by a table in the local code book. For example, a 2-inch pipe that connects to a building drain can carry four fixture units. A 3-inch pipe connecting to a building drain can accommodate up to 31 fixtures in this type of arrangement.

Island venting under this code is limited to sinks and lavatories. You must be careful when assessing this type of installation. Since the piping is done below the flood-level rim of the fixture, you must obey the rules for drainage piping. Vent-only fittings and pipe are not allowed in this type of setup.

Remember that I told you earlier that a vertical offset with two 45-degree fittings was considered a vertical pipe? Well, for this part of the code, forget it. Any offset in a waste stack vent is a violation of the code.

In this type of installation, every fixture drain must enter the waste stack separately. In other words, you cannot pick up multiple fixtures from branch drains. This is not all there is to the rules of a combination waste and vent system, but it's enough to show you the basics.

Another example

This is another example of how combination waste and vent systems can be regulated. These examples pertain primarily to southern states. The fixtures allowed for this system include sinks, dishwashers, floor sinks, waste receptors, floor drains, and similar fixtures.

This code warns its users to avoid situations that deliver too much water or sewage into a combination system. It further requires that commercial dishwashers and scullery sinks drain through a grease interceptor. These fixtures are also required to discharge into a floor sink through a minimum air gap.

Pipe sizes for this combination system are required to be twice as large as the pipes used under the normal, vented code. Horizontal fixture drains and above-ground P-traps are not required to meet this criteria.

Cleanouts are required in the vertical risers of these drains. The cleanouts must be installed near the tee fitting connecting the fixture to the drain.

Vents are required at the upper end of each drainage branch. These vents must be washed by the last fixture on the branch. Vents are also required where branches join. All fixtures must have vents located downstream.

The vertical distance between a fixture or drain outlet and a trap weir may not exceed 2 feet. Relief vents are required for long runs of piping.

As you can see, there are some substantial differences between the basics of these two codes. But we are not done yet. Let's move on to the code used in the western states.

A third example

Here is a third example of combination waste and vent systems. These rulings are based on one of the most expansive plumbing codes in the country. In this code, drain and vent pipes must be twice the size of the pipes used for drains in a standard plumbing installation. This code doesn't like to see any above-floor fixture, such as sinks and lavatories, piped in the combination method unless structural conditions demand it, such as island venting.

The goal of this combination is to serve the needs of floor drains and showers. Even so, the requirements for permission to use these methods should show hardship in installing the plumbing by conventional means. Commercial kitchen equipment is not recommended for combination systems.

Items, such as pumps, that produce high volumes or water, should not be connected to combination waste and vent systems under this code. The vertical length of risers should be kept at 24 inches or less. Relief vents are required on long runs of pipe. Trap sizes must also be oversized by a factor of 2.

This code spells out that the users of the information should seek advice from local code administrators if any doubt exists in the proper installation of a combination system.

There are similarities between the west and the south. The code used on much of the East Coast is a little more liberal, but wait until you see examples of what the state of Maine will allow. Keep in mind that none of the above codes allow toilets or urinals on their combination systems.

The final example of combination systems

The final example of combination systems will give you insight to what Maine's Alternate Plumbing code allows. I think you will be shocked to see the disparity between this code and the major codes previously discussed.

This code provision applies only to one- and two-story buildings, but for Maine, that covers most of the buildings. To qualify for these code applications, the buildings must not have plumbing fixtures, other than laundry hookups, in their basements.

All buildings must have a soil or waste stack, or stacks—as required by the provisions of the code—that extend in their full size through the roof of the building. The sizes of these pipes are determined with the use of a table provided in the code book.

Toilets are not allowed to discharge into pipes with diameters of less than 3 inches, but they are allowed on the combination waste and vent system. This in itself is a huge difference over the other codes. A 3-inch stack may carry the discharge of up to three toilets from any branch interval. A 3-inch stack is allowed to carry the waste of up to a total of six water closets—another major difference. Any building with a water closet in it must have at least one 3-inch stack terminating above the roof.

Vertical branches can extend up to 10 feet before an additional vent is required. Lateral branches can also extend 10 feet from their vertical vent. The vertical distance between a fixture trap and its drain may not exceed 30 inches.

Loop vents are required when a minimum of two and not more than eight toilets, pedestal urinals, slop sinks, or shower stalls are connected in a series.

Drum traps, which are illegal by most code standards, are expected in this system. Horizontal 3-inch drains are restricted to no more than one water closet if the grade is either ¼ or ½ inch to the foot, and they are prohibited if the grade is ⅛ inch to the foot or less.

It should be obvious to you by now that the rules, regulations, and laws in different codes can vary dramatically. You must always refer to your local code for the answers to your code questions.

Vent Materials

Vent materials are often the same as drain materials, but there are sometimes differences. Let me give you a list of the materials approved for vents by one of the major codes.

Above-ground vent materials

Above-ground vent materials (Table 17.4) could include ABS, PVC, brass, galvanized steel, cast iron, aluminum, copper, and lead. These materials may not be approved by all codes in all locations.

TABLE 17.4 Materials Commonly Approved for Above-Ground Vents

Cast iron
ABS
PVC
Copper
Galvanized
Lead
Aluminum
Brass

TABLE 17.5 Materials Normally Approved for Underground Vents

Cast iron
ABS
PVC
Copper

Below-ground vent materials

Below-ground vent materials (Table 17.5) could include ABS, PVC, cast iron, and copper. These materials may not be approved by all codes in all locations.

This has been a long chapter, but you should have learned more than most apprentices know about vents and venting. I hope you have developed a respect for the importance of vents in a plumbing system. Obviously, there are some extreme differences between the various codes; I trust you will consult your local code before engaging in the act of plumbing.

So far we have looked at underground plumbing, above-ground drainage piping, and venting procedures. When you turn this page, you will be embarking on an education in water distribution pipes. These are the pipes that bring potable (drinking) water to plumbing fixtures. Let's flip this page and see what is involved with the installation of water pipes.

Chapter

18

Water Distribution Systems

Water distribution systems are the systems that deliver potable (drinking) water to the plumbing fixtures in a building. This network of piping is very important. Without drinking water, we would all die. If the water delivered through the pipes were not potable, many forms of illness could result. The efficient delivery of potable water is critical to most plumbing systems.

The plumbing code can be very strict about the installation procedures for potable water piping, and it should be. The human dependency on good water is nothing to be taken lightly. Not only can installation procedures be regulated tightly, but the materials approved for the transportation of drinking water come under much scrutiny.

Even the materials used to join potable water pipes together gets a lot of attention. It was not very many years ago when the solder used to make joints in copper pipe was allowed to have a lead content of 50 percent. With the recent concerns over lead and its effect on the human body, new regulations have been installed to curb the use of solder that has a high content of lead.

It seems that as every year passes new rules come into play. These new regulations are the result of ongoing investigations into what it takes to provide a safe plumbing system. The whole purpose of the plumbing code is to provide safe and sanitary conditions for consumers. A handful of plumbers fight the regulations adopted by the codes, but in general, the code protects the health of the population.

Up to this point we have studied water services, sewers, underground plumbing, above-ground drainage systems, and plumbing vents. This chapter is going to delve into the code requirements and installation procedures for water distribution pipes.

Compared to the other topics we have covered, the requirements for water distribution pipes are simple, but that doesn't mean they are not important. Water distribution pipes and their proper installation are very important.

While many materials are approved for conveying potable water, the decision about what material to use is left up to the plumbing contractor. There are many times when one material will perform better than another. For example, if you were about to install the water pipes for a house and the well had a high concentration of acidic water, what type of pipe would you use?

Copper is an industry standard for water pipes. It normally gives years of trouble-free service, and it is easy for professional plumbers to work with. However, acidic water can destroy copper pipe and tubing. If a building's water supply is known to be acidic, polybutylene (PB) might be the best pipe material to install.

This chapter is going to show you what materials are approved for use and where they are put to their best use. Additionally, you will learn about many other aspects of the proper installation of a water distribution system. Now, let's get started on our journey into the world of water distribution pipes.

Isn't a Water Service a Water Distribution Pipe?

A water service is not a water distribution pipe. It does convey potable water to a building, but it is not considered a water distribution pipe. Water services fall into a category all of their own.

Water distribution pipes are the pipes found inside a building. Water services run the majority of their length outside of a building's foundation. They are usually buried in the ground. Water distribution pipes are installed within the foundation of the building and don't normally run underground; however, they can, such as in the case of a one-story building where the water distribution pipes are buried under a concrete floor. In any event, water services and water distribution pipes are dealt with differently by the plumbing codes.

Why Is Water Distribution Different from Water Services?

Water distribution pipes are dealt with differently than water service is because they serve different needs. Water distribution pipes pick up where water services leave off. The pipes distributing water to plumbing fixtures do not have to meet the same challenges that underground water services do.

For one thing, the water pressure on a water distribution pipe is often less than that on a water service. Most homes have water pressure ranges between 40 and 60 pounds per square inch. A water service from a city main could easily have an internal pressure of 80 pounds per square inch or more.

Water services are also buried in the ground. Water distribution pipes rarely are. The underground installation can affect the types of pipes that are approved for use. Another big difference between water service pipes and water distribution pipes is the temperature of the water they contain. Water services deliver cold water to a building. Water distribution pipes normally distribute hot as well as cold water. Since most codes require the same type of pipe to be used for the cold water lines as is used for hot water pipes, the temperature ranges can disqualify some types of materials. This is a significant factor when choosing a water distribution material.

In general, there are many reasons why a material approved for use for water service may not be approved for use as a water distribution pipe.

How Do You Decide What Materials to Use?

The first step in deciding what materials to use is to determine what materials are approved for use as water distribution pipes. This is easy enough; all you have to do is refer to your code book. We will look at the materials that are typically approved for water distribution in a moment. There are, however, other factors that may influence your decision about which material to use.

Copper tubing is probably the most common material used in modern water distribution systems. Is it the best? That is a argumentive question. Many plumbers consider PB to be a better water distribution pipe. The gray, flexible PB pipe does offer many advantages. It is easy to install, does not require any soldered joints, and it withstands extremely cold temperatures better than copper tubing. PB is also considerably less expensive than copper. Do these factors make it the best choice? Not necessarily. PB is a fine plumbing material, but there are reasons why many plumbers prefer copper. Before we get into the debate over which is better, let's look at the qualities of both materials from the viewpoint of the plumber in the field.

Working with copper

Many experienced plumbers prefer working with copper when building a water distribution system. One reason is probably that old habits are hard to break. Copper has been used for water distribution pipe for many, many years.

Before copper was used for water pipes, most systems were built with galvanized steel pipe. The galvanized pipe was heavy, time consuming to cut, needed to be threaded, and took substantial time to install. When copper started being used in place of galvanized steel, plumbers liked it. The copper tubing was lightweight, easy to cut, required no threads, and went together quickly. There were many advantages to using copper as a water distribution material.

I didn't enter the trade until copper was already the predominant choice for water distribution pipes, but I am willing to bet there were plumbers back then who didn't want to make the switch from galvanized pipe to copper tubing. They probably thought the thin-walled copper wouldn't hold up as well as steel pipe, and they probably argued that the threaded joints of the steel pipe were better than any soldered joint. I suspect the old debate between copper and galvanized pipe was as much an issue, for its time, as the present-day controversy between copper and polybutylene.

It is often said that history repeats itself. Could today's question over which is better, copper or PB, be a sign of history repeating itself? Take a look at the comparisons. In the old days, copper weighed less than galvanized steel pipe. Today, PB weighs less than copper tubing. Galvanized pipe was an accepted standard as a water pipe material in its day. Copper is an accepted standard for water pipe today. In the old debate copper was faster and easier to work with than galvanized pipe. Polybutylene is now faster and easier to work with then copper. Can you see history repeating itself?

Copper is still used extensively as a water distribution material, but PB is gaining popularity. What makes the coiled pipe so popular? Well, let's have a look.

The popularity of PB pipe

The popularity of PB pipe has grown in recent years. The material allows plumbers to do work in ways they never could with copper. PB is extremely flexible, allowing it to be installed with methods similar to running electrical wiring. The pipe can be snaked up walls, like a wire. This is a big advantage to plumbers working on remodeling jobs.

PB pipe can expand a great deal before it splits. This gives it an advantage over copper in cold climates. When copper tubing would split from having its contents frozen, PB would simply expand. The pipe can, of course, expand to the point of splitting, but it will not split from freezing temperatures as quickly as copper.

There are no joints to solder when working with PB. Insert fittings and special crimp rings are used to make the joints. This is something of an advantage, but it can also be a disadvantage. There are times when getting the crimping tool into position to make a joint is more difficult than it would be to put a torch in the same space.

PB comes in coils of continuous pipe. This allows long runs of pipe without the need for couplings. Rigid copper is sold in 10- and 20-foot lengths. Runs of more than 20 feet require couplings. Every joint in a pipe is a potential leak; the fewer joints there are, the less chance there is for trouble.

Soft copper is available in coils, usually about 60 feet long. However, soft copper is expensive, and it is rarely used for substantial portions of a water distribution system. While we are talking money, lets compare copper to PB.

The following prices are based on advertised pricing from one of my current suppliers. The price per foot for ¾-inch coiled PB pipe is $0.39. A crimp ring for making a joint is $0.13. An elbow fitting for the pipe is $0.67, and a tee fitting is $0.86.

If we look at same materials made from copper, the prices are a little different. The price per foot for ¾-inch type-L rigid copper is $1.34. There is no crimp ring required for copper joints, but solder, flux, and torch fuel are a factor. A ¾-inch copper elbow fitting is priced at $0.37, and a tee costs $0.78.

Copper costs almost a dollar a foot more than PB. However, PB fittings are more expensive, and crimp rings are required. Generally speaking, PB is the less expensive of the two materials when plumbing a water distribution system.

As you can see, there are many good reasons to use PB as a water distribution pipe. The decision about what type of material to use is a personal one, so long as the material is approved for the use.

Materials Approved for Water Distribution

There are a number of materials approved for water distribution, but only a few are used frequently in new plumbing systems. While I don't have any formal statistics in front of me, I would guess that copper is still the number 1 choice as a water distribution material. I would think PB pipe comes in second, and that chlorinated polyvinyl chloride (CPVC) runs a distant third.

Materials approved for use in water distribution systems may vary slightly from code to code. The information I am about to give you is valid in many jurisdictions, but remember to always check your local code before installing plumbing.

Galvanized steel pipe

Galvanized steel pipe is an approved water distribution pipe. It has been used for a very long time, and you will still find it in many older plumbing systems. It is not, however, used much in modern plumbing.

This pipe is heavy and requires threaded joints. The pipe tends to rust and deteriorate more quickly than other approved water distribution pipes. There are few, if any, occasions when galvanized steel pipe is used for modern plumbing installations.

Brass pipe

Brass pipe is approved for water distribution, but like galvanized pipe, it is rarely used.

Copper

We have spent considerable time talking about the use of copper as a water distribution pipe; it is, of course, approved for that application. There are, however, different types of copper. By types, I mean ratings of thickness. The two types most commonly used for water distribution are type-L and type-M. Type-K copper is also fine for water distribution, but it is more expensive and usually the extra-thick pipe is not needed. Type-M copper, the thinnest of the three, is not approved in all areas for water distribution. For years type-M was the standard, but now type-L is required by some administrative authorities.

Polybutylene

PB is approved for water distribution.

CPVC

CPVC is approved for use as a water distribution material. This rigid plastic pipe is popular with do-it-yourselfers, but most professional plumbers prefer not to use it.

CPVC gets brittle in cold weather, and its joints tend to leak more often than other types of pipes. A cleaner and primer should be used on the pipe before it is glued into its fittings. If the joint is disturbed before it drys, leaks are likely. Working with this pipe takes too much time for most professionals. The installation of copper or PB goes much faster for the pros.

What Purposes Require Potable Water?

The important needs for potable water include drinking, cooking, bathing, and the preparation of food and medicine. Potable water is also used for other activities, but the ones listed are the uses dealt with in the plumbing code. Fixtures used for any of the above purposes must be plumbed in a way that only potable water is accessible to them.

What Are the Requirements for Providing Hot Water?

Hot water must be provided in buildings where people work and for all permanent residences.

How About Cold Water?

Cold water must be supplied to every building that contains plumbing fixtures and is used for human occupancy.

How Much Water Pressure Is Required?

The water pressure for a water distribution system must be high enough to provide proper flow rates at each of the fixtures. Flow rates for various fixtures are determined by referring to the tables or text in your local code.

Generally, 40 pounds per square inch is considered adequate water pressure. If the incoming pressure is 80 pounds per square inch or more, the pressure must be controlled with a pressure-reducing valve. This valve is installed between the water service and the main water distribution pipe. The device allows the water pressure to be regulated to a lower pressure rating. There are, of course, exceptions to most rules; check your local code for the requirements in your jurisdiction.

Water Conservation

Water conservation is an issue that most codes deal with. The codes restrict the flow rates of certain fixtures, such as showers, sinks, and lavatories. An unmodified shower head can normally produce a flow of 5 gallons per minute. This flow rate is often reduced to 3 gallons per minute with the insertion of a water-saver device. The device is a small disk with holes in it.

Other water-saver regulations apply to buildings with public restrooms. It is frequently required that all public-use lavatories be equipped with self-closing faucets. These faucets are restricted in their flow rates, and they cut themselves off after use.

Urinals are supposed to have flow rates of not more than 1½ gallons per flush. Toilets generally are not allowed to use more than 4 gallons of water during their flush cycle.

Antiscald Devices

Antiscald devices are required on some plumbing fixtures. Different codes require these valves on different types of fixtures, and they sometimes set different limits for the maximum water temperature. Antiscald devices are valves or faucets that are specially equipped to avoid burns from hot water. These devices come in different configurations, but they all have the same goal: to avoid scalding.

Whenever a gang shower is installed, such as in a school gym, antiscald shower valves should be installed. Some codes require antiscald valves on residential showers, but others don't. The maximum temperature allowed for hot water in some codes is 110 degrees F. Other codes set the high-temperature limit at 120 degrees F.

Valve Locations

There are several mandatory valve locations in a water distribution system. The types of valves placed in these locations is also mandated by the code. For example, a full-open valve is required near the origination of a water service. In the case of a city water hookup, the valve is supposed to be installed near the curb of the street. Water services extending from a well are not required to have such a valve at their origination. For those of you who don't know, a full-open valve is a valve like a gate valve or ball valve. When these valves are open, water has the full diameter of the valve to flow through, rather than a restricted opening like those in valves with washers and seats.

A full-open valve is required on the discharge side of every water meter. The main water distribution pipe must be fitted with a full-open valve where it meets the water service. In nonresidential buildings, a full-open valve must be installed at the origination of every riser pipe. In these nonresidential buildings, a full-open valve is required at the top of all drop-feed pipes.

If a water supply pipe is feeding a water tank, such as a well pressure tank, a full-open valve is required. All water supplies to water heaters must be equipped with full-open valves. If a building contains multiple dwellings, the water supplies for each dwelling must be equipped with full-open valves. I know this disclaimer is starting to get old, but be aware that not all codes have the same requirements. Check your local code for applicable restrictions. Since I assume by now that you know you have to verify permissible plumbing standards with the use of your local code, I am going to stop reminding you.

The valves we are about to discuss are not required to be full-open valves. The valves used in the following locations are usually either stop and waste valves (Fig. 18.1) or supply-stop valves (Fig. 18.2). Valves are required on all supply pipes feeding silcocks and hose bibbs. If a water supply is serving an appliance, such as an ice maker or dishwasher, a valve is required.

Some types of buildings are required to have individual cutoffs on the

Figure 18.1 Stop and waste valve.

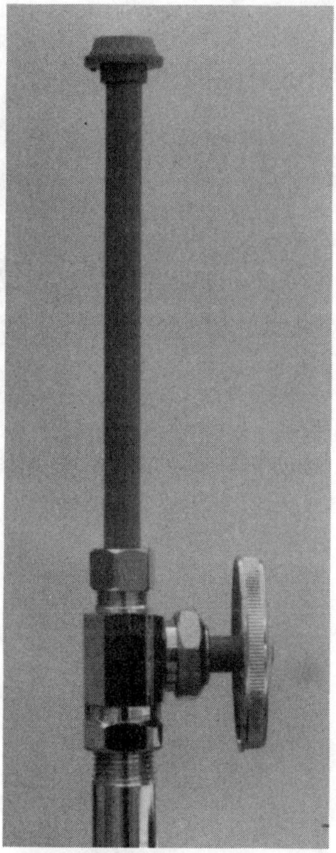

Figure 18.2 Straight stop fixture valve.

supplies to all plumbing fixtures. Other types of buildings are not. It is generally standard procedure to install cutoffs on all water supplies, with the possible exception of the pipes feeding tub and shower valves.

When a valve is required by the code, that valve must be accessible. It may be concealed by a door that opens or some other means of accessible concealment, but it must be accessible. Stop and waste valves are normally not approved for use below ground.

Pipe Support

Pipe support is always an important issue, and apprentices are often the ones responsible for it. If you look in your code book, you will find either text or a table to guide you in the placement of pipe hangers. There are many ways to secure water pipes (Figs. 18.3, 18.4, and 18.5).

Figure 18.3 Clip hanger securing horizontal pipe.

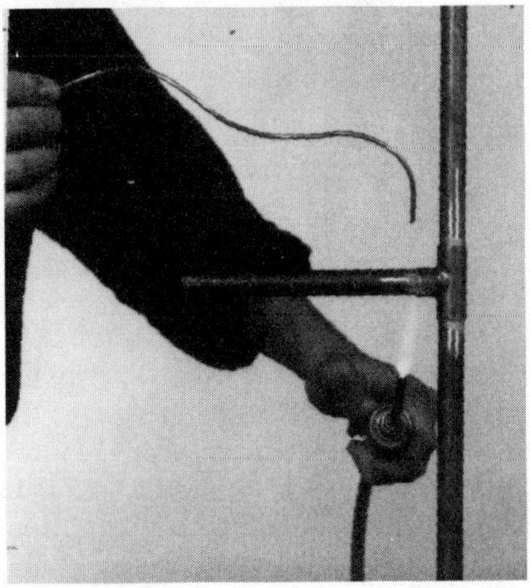

Figure 18.4 Clip hangers securing vertical pipes.

The distance between the hangers will vary, depending upon the type of pipe or tubing being used. Let me give you some common examples.

Copper tubing with a diameter of 1¼ inches or less should be supported at intervals not to exceed 6 feet. Larger copper can run 10 feet between supports. These recommendations are for pipes installed horizontally. If the pipes are run vertically, both sizes should be supported at intervals of no more than 10 feet.

If your material of choice is CPVC, it should be supported every 3 feet, whether installed vertically or horizontally.

PB pipe that is installed vertically is required to be supported at in-

Figure 18.5 Drive hanger securing horizontal pipe.

tervals of 4 feet. When installed horizontally, PB pipe should have a support every 32 inches.

Hot Water

Hot water will be provided by most water distribution systems, and there are some special rules that apply to the delivery of hot water (Fig. 18.6). Let's take a look at them.

All water heaters must be of an approved type. The heaters must be large enough to provide adequate hot water as established by the tables and text found in local code books. They must meet both the needs for daily requirements and for peak usage.

Water heaters must have a minimum pressure rating of 125 pounds per square inch, and the pressure rating must be marked permanently on the water heater in an accessible location. Every water heater must be equipped with a temperature and pressure relief valve (Fig. 18.7). The ratings for temperature and pressure valves will be determined by the pressure ratings on the water heaters. However, maximum temperature ratings should not exceed 210 degrees F, and maximum pressure ratings should not exceed 150 pounds per square inch. These valves are required to monitor the top 6 inches of water in the tank. The discharge from a relief valve must be piped downward to a point of safe discharge.

The discharge pipe may be piped to the sanitary drainage system with the use of an indirect connection. For example, the discharge pipe could terminate above a floor drain, but it may not be piped into the drainage system with a closed fitting.

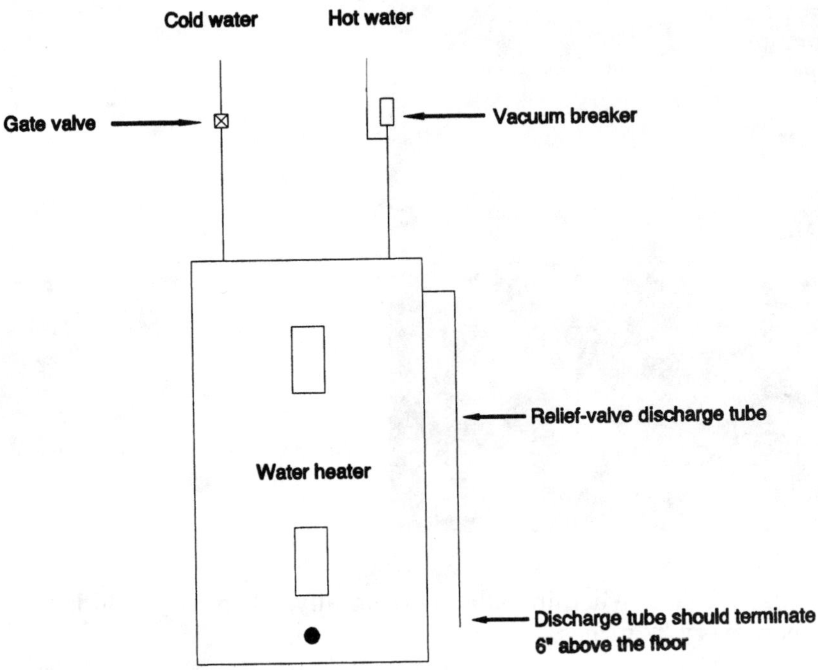

Figure 18.6 Typical water heater setup.

Discharge pipes are not allowed to have threaded ends at their point of termination. If they were, someone might install a device that would prevent the proper functioning of the pipe.

The discharge pipes from water heaters are not required to be plumbed into the sanitary drainage system. These pipes may terminate above a floor or at any other point where they do not present a hazard. When stopped above a floor, discharge pipes are commonly held about 6 inches above the floor level.

When water heaters are installed above habitable space, they may be required to sit in a pan or other container that will protect the living space from flooding should a problem occur. In such cases, there will be a drain outlet in the retaining pan that should be piped to an appropriate place for the disposal of unwanted water.

Every water heater should be equipped with a drain at its lowest point. Even if the tank is only a hot water storage tank, it should have a drain.

Water heaters are typically required to meet minimum standards for energy efficiency. There are requirements governing the amount of insulation required for water heaters.

The controls for adjusting and maintaining the temperature of hot

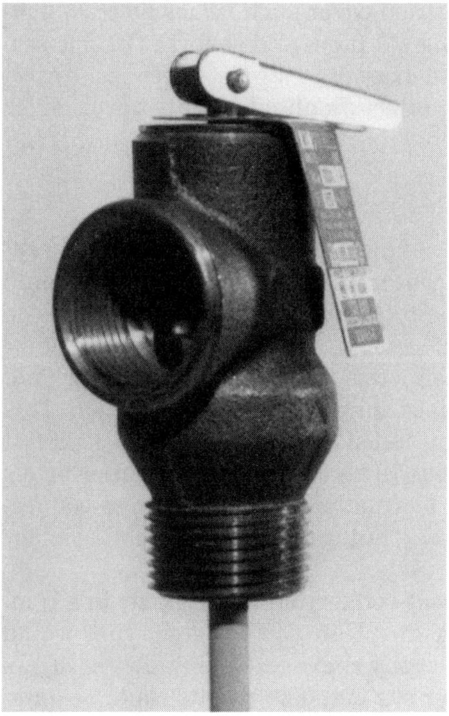

Figure 18.7 Relief valve.

water must be accessible. The temperature controls should function automatically to maintain the desired temperature.

Valves should not be installed on the outgoing hot water pipe from a water heater and valves should not be installed on the discharge pipes of relief valves.

If the distance between a water heater and a plumbing fixture receiving hot water exceeds 100 feet, steps must be taken to maintain hot water within 100 feet of the fixture. This might be accomplished by insulating the hot water pipe, but a recirculating system will probably be required. When recirculating pumps are used, they must be equipped with a switch that allows them to be cut off when not in use.

Electric water heaters must have a separate switch controlling their power source. If the water heater is fueled by oil or gas, a cutoff valve must be installed on the fuel supply line.

Protecting Potable Water Systems

Protecting potable water systems is an issue all codes address. They define their requirements differently at times, but they all agree protection is needed. The forms of protection cover such issues as cross-connections, backflow, vacuums, using old pipe, and so forth. In this section of the chapter we are going to examine many of these issues and make you aware of what types of protection plumbers must provide for potable water systems.

Keeping it clean

One of the most important aspects of protecting potable water is keeping it clean. If the water becomes contaminated, health problems could arise. One common way contaminants can invade a potable water system is through a cross-connection.

There are many types of cross-connections possible. For example, hot water may pass into cold water pipes through a cross-connection. This shouldn't contaminate the system, but it can create a health risk. If someone turns on what they believe to be cold water and discover it's hot water, they could be burned. Normally, this type of cross-connection is more a nuisance than a threat, but it does have the potential for producing injuries.

A much more serious type of cross-connection might occur in a commercial photography processing plant. Can you imagine what would happen if a pipe conveying photography chemicals were allowed to mix its contents with the potable water system? The results could be quite serious indeed.

Cross-connection are normally created by accident, but sometimes they are designed. If a cross-connection of some sort is to be installed, it must be protected with the proper devices to avoid unwanted mixing. Few, if any, codes will allow a cross-connection between a private water source, such as a well, and a municipal water supply.

Care must be taken when installing plumbing not to create unintentional cross-connections. I mentioned earlier in the book how a plumber might confuse a heating pipe with a water distribution pipe. That was a good example of an unwanted cross-connection. If only approved materials are used and if they are installed properly, cross-connections should not occur.

Cross-connections are not the only way that a potable water system might become contaminated. The use of old pipe could be a source of contamination. If old pipe has been used for any purpose other than the conveyance of potable water, it should not be used in a potable water system.

If contaminants get into a private water source, such as a well, the potable water system can become fouled. Wells must be covered and protected against the entry of nonpotable water. This is why the hole in the side of the well casing for the water service must be sealed.

Water storage tanks are another possibility when it comes to ruining a potable water system. If the interior of the tank is painted or otherwise covered with a material that changes the condition of potable water, a system could be contaminated. There are other ways a potable water system could be damaged. Let's see how backflow and vacuums might wreak havoc on a potable water supply system.

Vacuums and Backflow

Vacuums and backflow can cause a potable water system to become contaminated. To reduce the risk of this happening, plumbing codes require certain protective devices to be installed in specific locations. All potable water systems must be protected against backflow.

Air gap

An air gap is the open air space between the flood-level rim of a fixture and the bottom of a potable water outlet. For example, there is an air gap between the outlet of a spout on a bathtub and the flood-level rim of the tub, or at least there is supposed to be.

By having such an air gap, nonpotable water, such as dirty bath water, cannot be sucked back into the plumbing system through the potable water outlet. If the tub spout was mounted in a way that it was submerged in the tub water, the dirty water might be pulled into the water pipes if a vacuum was formed in the plumbing system.

The amount of air gap required on various fixtures is determined by the plumbing code. For example, you might find in a table or in the text of your code that a tub spout must not be installed in a way to make its opening closer than 2 inches to the flood-level rim. The same ruling for lavatories might require the spout of the faucet to remain at least 1 inch about the flood-level rim.

There is another type of air gap (Fig. 18.8) used in plumbing. This type is a fitting that is used to accept the waste from a dishwasher for conveyance to the sanitary drainage system. The device allows the water draining out of the dishwasher to pass through open air space on its way to the drainage system. Since the water passes through open air, it is impossible for waste water from the drainage system to be siphoned back into the dishwasher.

Backflow preventers

Backflow preventers (Fig. 18.9) are devices that prevent the contents of a pipe from flowing in a direction opposite of its intended flow. These protection devices come in many shapes and sizes, and costs range from a few dollars to thousands of dollars. Backflow preventers must be installed in accessible locations.

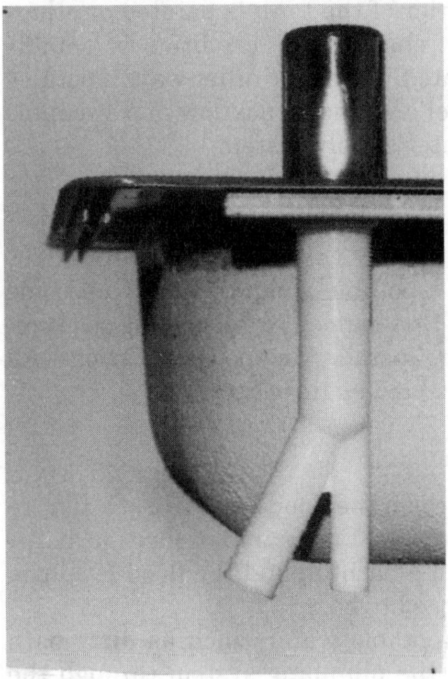

Figure 18.8 Air gap.

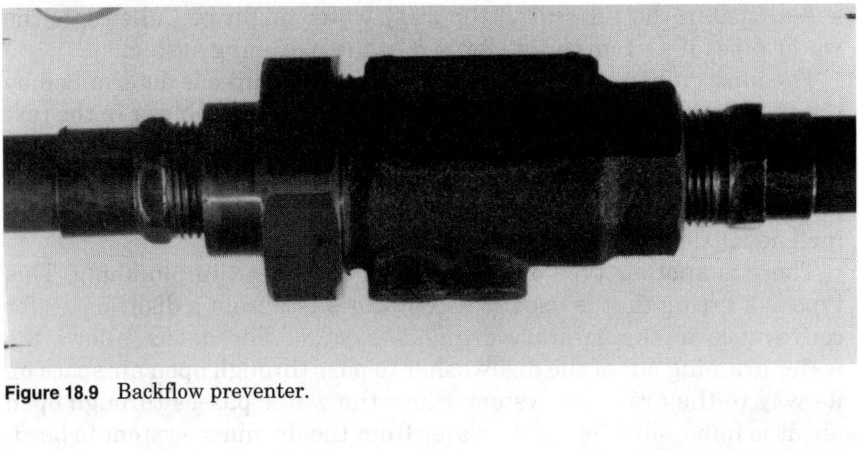

Figure 18.9 Backflow preventer.

When a potable water pipe is connected to a boiler to provide it with a supply of water, the potable water pipe must be equipped with a backflow preventer. If the boiler has chemicals in it, such as antifreeze, the connection must be made through a reduced-pressure-principle backflow preventer of an air gap.

Whenever a connection is subject to back pressure, the connection must be protected with a reduced-pressure-principle backflow preventer.

Check valves

Check valves have an integral section, something of a flap, that allows liquids to flow freely in one direction and not at all in the opposite direction. Check valves are required on potable water supply pipes that are connected to automatic fire sprinkler systems and standpipes.

Vacuum breakers

Vacuum breakers (Fig. 18.10) are installed to prevent a vacuum being formed in a plumbing system that could result in contamination. Some of these devices mount in-line on water distribution pipes and others screw onto the threads of silcocks and laundry-tray faucets.

When vacuum breakers are required, the critical level of the device is usually required to be set at least 6 inches above the flood-level rim of the fixture it is serving. The critical level is the point at which the vacuum breaker would be submerged, prior to backflow.

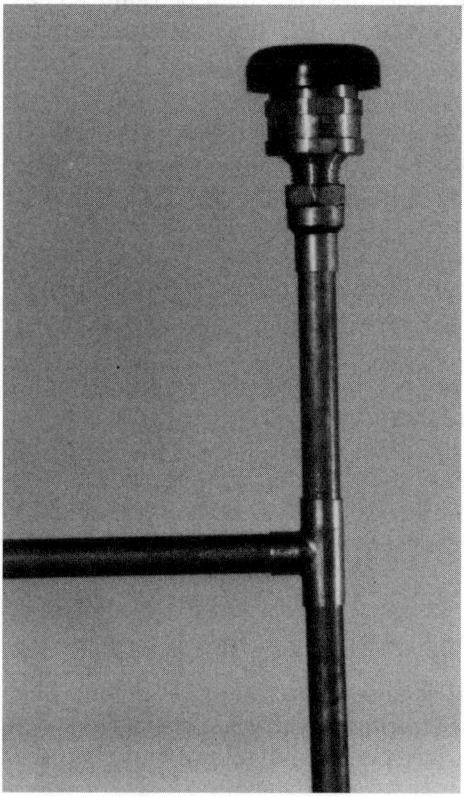

Figure 18.10 Vacuum breaker.

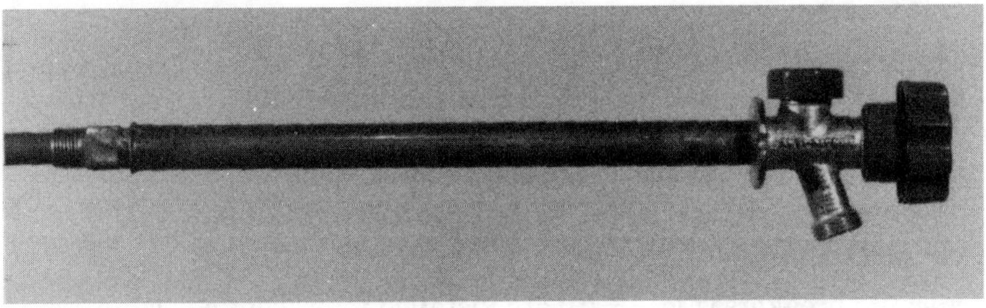

Figure 18.11 Frost-free silcock with built-in vacuum breaker.

Normally, any type of fixture that is equipped with threads for a garden hose must be fitted with a vacuum breaker. Some silcocks are available with integral vacuum breakers (Fig. 18.11). For the ones that are not, small vacuum breakers that screw onto the hose threads are available. After the devices are secured on the threads, a set screw is turned into the side of the device. The screw is tightened until its head breaks off. This ensures that the safety device cannot be removed easily. Remember the woman that was spraying her garden when her water heater died? If her silcock had been equipped with a vacuum breaker, her pipes would not have been contaminated.

Since vacuum breakers have an opening that leads to the potable water system, they cannot be installed where toxic fumes or vapors may enter the plumbing system through the openings. One example of a place where they could not be installed is under a range hood.

There are other circumstances when some form of backflow protection is required, but we have covered most of the common situations.

More Information

We are not done yet, there is still more information for me to give you on potable water systems. Some of the topics left to cover include sizing water pipes, installing water pipes, and personal pointers from a master plumber's point of view. If you need to take a break from reading, do it now. You won't want your concentration broken once we get into the remaining issues.

Sizing water pipes

The task of sizing water pipes strikes fear into many plumbers. They take one look at the friction-loss charts in their code books, see some math formulas that read like a foreign language, and run the other way. Sizing water pipes can be intimidating. Fortunately for you, apprentices don't have to size water pipes. However, to pass your plumbing exam, you will have to learn something about sizing somewhere along the way. I can't think of a better place for you to get introduced to sizing techniques than right here, in this book.

If you are thinking about skimming—or worse yet, skipping—this section, don't. I promise you that the way I'm going to show you to size water pipes will not make you pull your hair out. Now, if you're ready, let's get started.

There are two approaches to sizing water pipe: the engineer's approach and the plumber's approach. Large jobs typically must to be designed by an engineer or other suitable professionals, not plumbers. Most plumbing codes will not allow plumbers to design complex water distribution systems. For most plumbers, myself included, this is a blessing.

Your local code book will provide information about how you can use formulas, friction charts, and other complicated aids to size water pipes, but they will also give you some easier ways around the problem.

The first question an inexperienced apprentice is likely to have is how to figure out what size pipe is required for supplying water to various fixtures. This is a valid question, and you will be happy to know it is an easy one to answer. By referring to the text and tables in your code book, you will be able to size fixture supplies immediately (Tables 18.1 and 18.2).

The information in your code book will make it clear what size pipe is needed for all common plumbing fixtures. Let me give you some idea of what you will find. One of the major codes allows fixture supplies with diameters of ⅜ inch to be used for the following fixtures: bidets, drinking fountains, single showers, lavatories, and water closets—with

TABLE 18.1 Required Capacities at Fixture Supply Outlets

Fixture	Flow rate, in gallons per minute	Flow pressure, square inch in pounds per
Bathtub	4.00	8
Bidet	2.00	4
Dishwasher	2.75	8
Hose bibb	5.00	8
Kitchen sink	2.50	8
Laundry tub	4.00	8
Lavatory	2.00	8
Shower	3.00	8
Water closet (two pieces)	3.00	8
Water closet (one piece)	6.00	20

TABLE 18.2 Common Minimum Fixture-Supply Sizes

Fixture	Minimum supply size, in inches
Bathtub	½
Bidet	⅜
Shower	¾
Toilet	⅜
Lavatory	⅜
Kitchen sink	½
Dishwasher	½
Laundry tub	½
Hose bibb	½

the exception of one-piece toilets. If you increase the fixture supply size to ½ inch, you can feed these fixtures: bathtubs, domestic dishwashers, hose bibbs, residential kitchen sinks, laundry trays, service sinks, flush tank urinals, wall hydrants, and one-piece toilets.

By increasing the fixture size to ¾ inch, you can connect to commercial kitchen sinks, flushing rim sinks, and flush valve urinals. Stepping up to a 1-inch pipe, you can serve toilets that use flush valves.

Now wasn't that easy? All you have to do is look at your code book, and you can tell quickly what the minimum size requirements are for your fixture water supplies. Up to the first 30 inches of a section of pipe connecting to a fixture is considered a fixture supply. Once the supply is more than 30 inches away from the fixture, it is considered a water distribution pipe. If you want to know what the required flow rates and pressures for the various fixtures are, you can simply refer to your code book. All of this information is available to you in an easy-to-understand format.

The next step is not quite as easy, but it still is not difficult. There are several methods for sizing water pipe. Most code books disclaim any responsibility for the examples they give on sizing. The examples are just that—examples, not carved-in-stone procedures. One of the easiest ways to size water pipe is with the use of the fixture-unit method, and that is the one we are going to concentrate on. Let me show you how it works.

The fixture-unit method

The fixture-unit method of sizing water distribution pipes is pretty easy to understand. You will need to know the total number of fixture units that will be placed on your pipes. This information can be obtained from listings in your code book.

The total developed length of your water pipes will have to be known. If you are on the job, you can measure this distance with a tape measure. When you are working from blueprints, a scale rule will give you the numbers you need. The measurement will begin at the location of the water meter.

You will also need to know what the water pressure on the system will be. If you are not controlling the water pressure, as you might with a pressure-reducing valve or a well system, you can call the municipal water department to obtain the pressure rating on the water main.

To pinpoint the accuracy of your sizing, you should learn the difference in elevation between the water meter and the highest plumbing fixture. However, by being generous with your pipe sizes, you can get by without this information most of the time.

The fixture units assigned to fixtures for the purpose of sizing water pipe will normally include both the hot and cold water demand of the fixture. The ratings for fixtures used by the public will be different from those installed for private use. All of the information you need for this method of sizing is easily obtained from you local code book.

Now, lets assume we are going to size the water pipe for a small house. The house has one full bathroom and a kitchen. It also has a laundry hookup and one hose bibb. We need to know what fixture-unit ratings to assign to these fixtures. A quick look at a code book might give us the following ratings:

Fixture	No. of fixture units
Bathtub	2
Toilet	3
Hose bibb	3
Lavatory	1
Sink	2
Laundry hookup	2

Okay, now we know our total number of fixture units. The working pressure on this water system will be 50 pounds per square inch. Our total developed length of piping, from the water meter to the most remote fixture, is 95 feet.

We now have all we really need to size our water pipe, with a little help from a table in our code book. When we refer to our sizing table in the code book, we will see different tables for different pressure ratings. We choose the table that matches our rating of 50 pounds per square inch.

There will be columns of numbers to identify fixture units and developed lengths of pipe. We will see sizes for building water supplies and branches. Even the sizes for water meters will be available. All we have to do is put our plan into action.

Our water meter for this example is a ¾-inch meter. First we have to size the water service. How many fixture units will be on the water service? When you count all our fixture units, the total is 13. Our pipe has to run 95 feet, so what size does it have to be? By looking at the sizing table in your code book, you will see quickly what size it must be. In the case of this example, the required size is a 3/4-inch water service. I was able to determine this just by running my finger down and across the sizing table.

The sizing table is based on the length of the pipe run, the water pressure, and the size of the water meter. Knowing those three variables, you can size pipe quickly and easily. The same cross-referencing that was done to size the water service also works when sizing water distribution pipes.

For example, assume that the most remote fixture in the house is 40 feet from the end of the water service. When I look at my table, I see that a 3/4-inch pipe is required to begin my water distribution system. As the main 3/4-inch pipe reaches fixtures and lowers the number of fixture units remaining in the run, the pipe size can be reduced. As a rule-of-thumb, when there are only two fixtures remaining to be served on a residential water main, the pipe size can drop down to a 1/2-inch pipe.

Let's say my 3/4-inch pipe is running down the center of the home. The kitchen sink is perpendicular to the main water line and lies 22 feet away. The sink has a fixture-unit demand of two. By checking my sizing table, I see that the pipe branching off the main to serve the sink can be a 1/2-inch pipe.

Once you sit down with your code book and work with the fixture-unit method of pipe sizing, you will see how easy it is. Now, we are done with our sizing exercise, and we are going to move on to some techniques used to install water distribution pipes.

Putting It All Together

We've talked a lot about water distribution systems; now let's talk about putting it all together. The three main types of water distribution materials used by professionals are copper, PB, and CPVC. Each of these types of materials requires a different technique for their installation. Let's look at each type of material and see how joints are made.

Copper

Copper tubing used for water distribution systems is usually put together with soldered joints. The need for soldering skills when installing copper is what sends most do-it-yourselfers running to the hardware store for CPVC pipe. However, soldering skills are not difficult to master.

Most copper water pipe has a diameter of 1 inch or less. Much of it is either 3/4- or 1/2-inch pipe. These small pipes, or tubing as they should be called, are simple to solder.

The first step may involve cutting the pipe. This is best done with a roller-type cutter. Once the pipe is cut, it should be cleaned. This is done by rubbing sandpaper on the outside end of the pipe. Be careful not to let slivers of copper puncture your hand. If the pipe was cut poorly, there may be jagged edges of sharp copper.

Shine the end of the pipe until you have a shiny band all the way around the pipe (Fig. 18.12). The band should be about 1 inch wide. Using a cleaning brush—a special wire brush sized to fit snugly into copper fittings—clean the inside hub of the copper fitting, turning the brush clockwise.

You now have a shiny pipe and fitting. Using a flux brush—a small brush made for applying flux—dip into your can of flux and apply flux to the inside hub of the fitting and to the end of your pipe (Fig. 18.13). Flux is a special compound used to clean copper and to aid in making a good solder joint. Without flux, solder will not run around the joint very well. Be careful not to get the flux in your eyes or in an open cut; it burns. It also doesn't taste very good, so keep it out of your mouth.

When the pipe and fitting are well fluxed, slide the pipe into the fitting; make sure it goes all the way in. You are now ready to solder your joint. You should be wearing safety glasses, long sleeves, and perhaps gloves. Dripping solder is hot, and it hurts when it lands on bare skin.

Get your solder, and remember, it should be lead-free. Check your local code for approved soldering contents in your area, but I'm sure

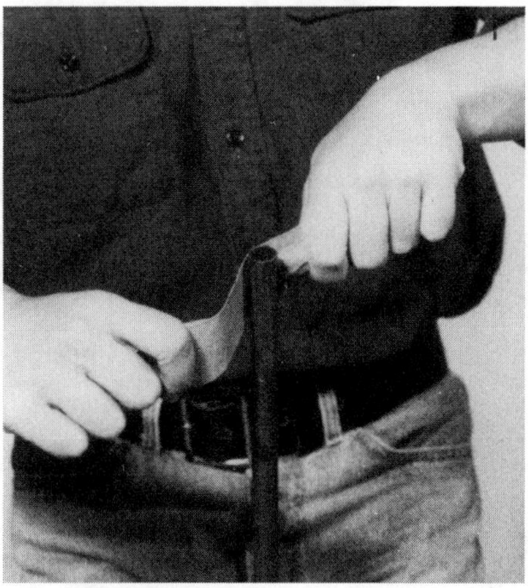

Figure 18.12 Sanding copper tubing.

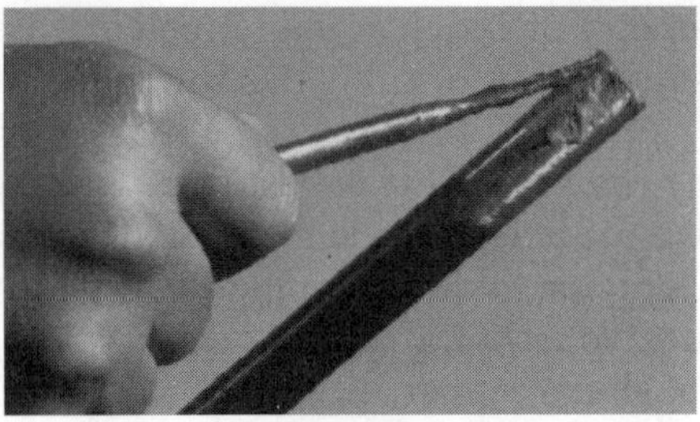

Figure 18.13 Applying flux to copper tubing.

you will find the lead content must be very low. Check your torch connections and cut on the valve. Light the torch with a striker.

What happens next will take a little hands-on experience to master. Hold the flame of the torch on the hub of the copper fitting (Fig. 18.14). Small pipes heat up fast, and larger ones take longer. If you hold the torch on the fitting too long, the fitting will turn black and the joint will probably leak. Not getting the fitting hot enough will result in lumping solder and a leak.

As the torch is heating the fitting, touch the tip of the solder to the point where the pipe enters the fitting. Periodic touching of the solder to the copper will tell you when the joint is ready. When the temperature is right, the solder will melt and run around the pipe, being sucked into the fitting. Done correctly, the joint will not leak. Experienced hands can make solder joints so cleanly that the solder is barely visible. Rookies usually leave an ample coating of solder on the outside of

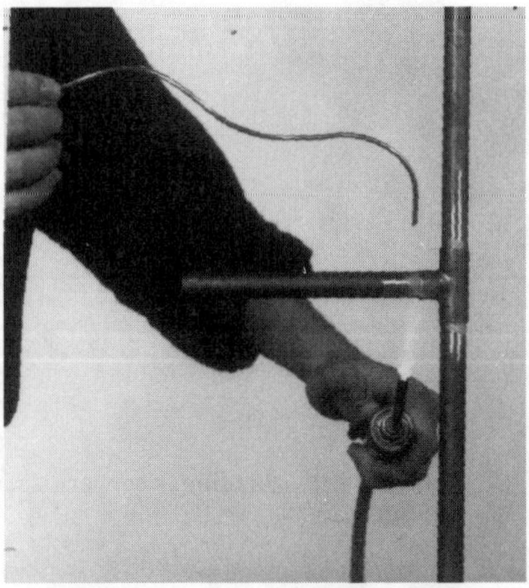

Figure 18.14 Soldering a horizontal joint.

the pipe, but that's okay when you are learning.

Once the joint is made, don't wiggle it. The solder needs time to cool before it can be moved without causing a leak. Fittings made from heavier metal, like brass valves, take longer to heat up than thin copper fittings. Once you have given the joint a few seconds to cool, wipe the excess flux from the joint with a rag. If you leave the flux on the pipe and fitting, it will turn green and ugly. Inspectors and consumers don't like to see that kind of sloppy work.

Learning to solder horizontal joints is pretty easy, but vertical joints intimidate some trainees. If you are soldering a vertical joint, hold the flame of the torch in a position to pull the solder into the joint (Fig. 18.15). For example, assume you are soldering a coupling in a riser. Solder the top section of the coupling first, and hold the flame on the middle of the fitting. When the top joint is made, keep the flame on the center of the coupling, and the solder on the bottom joint should be pulled up into the fitting. As I said, soldering requires some practice to master.

Flare joints

Flare joints are very rarely used in water distribution, but you might find some in older mobile homes. These joints are made with a flaring tool.

When copper tubing is put together with flare joints, the pipe must be rolled back to create the flare. To do this, you will need a flaring tool. Some larger pipes are flared open with tools that are driven into their ends, but most flare joints are made with a flaring block.

Remove the flare nut from the flare fitting you will be connecting the pipe to. Slide the nut over your pipe. Put the end of the pipe into the

Figure 18.15 Soldering a vertical joint.

Figure 18.16 Putting copper tubing in a flaring block.

flaring block and lock the block closed (Fig. 18.16). Position the flaring device over the center of the pipe and turn the handle clockwise (Fig. 18.17). As the flaring device enters the end of the pipe, the copper will be rolled back. Once the handle will not turn any more, loosen it, and remove your pipe.

The flared pipe will butt into the bevel on the flare fitting (Fig. 18.18). The flare nut will slide up to the flared end and screw onto the fitting. That's all there is to flaring. However, learning how much pipe to leave sticking up in the flaring block will take a little practice.

Compression joints

Making compression joints with copper tubing is a piece of cake. Remove the compression nuts and slide them over the ends of your pipes. Next, slide the compression sleeves over the pipe ends. Insert the pipe ends into the compression fitting and secure the compression nuts. As the nuts are tightened, the compression sleeves seal the joint. These joints are quick and easy to make.

Polybutylene

Making joints with PB is simple when you have the proper tools. You will need insert fittings, crimp rings, and a crimping tool. To make these joints, slide a crimp ring over the end of the pipe. Push the insert fitting into the end of the pipe. Slide the crimp ring into position over

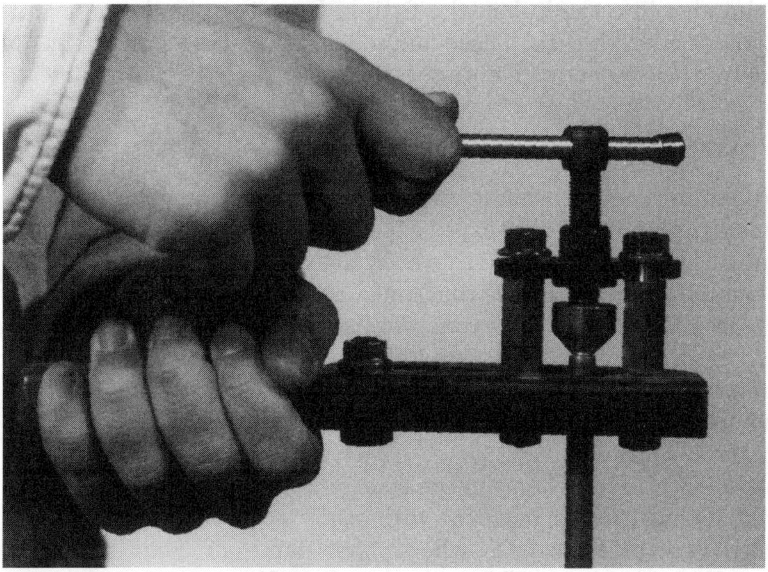

Figure 18.17 Flaring copper tubing with a flaring tool.

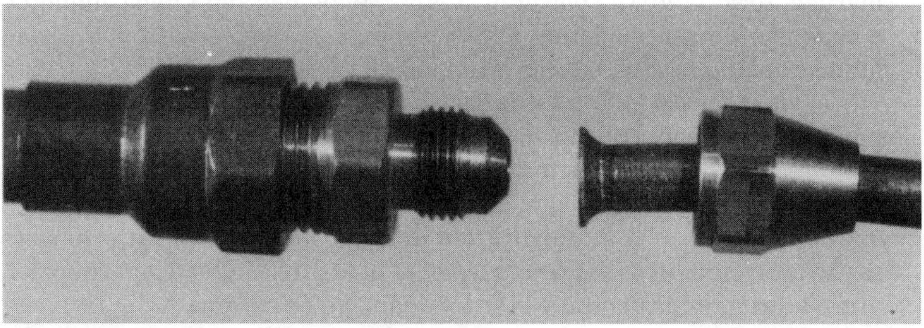

Figure 18.18 Making a flare connection.

the ridges of the insert. Open the jaws of the crimping tool, place it over the crimp ring, and squeeze. When the crimping tool's handles come together, your joint is made.

CPVC

CPVC is worked with much like PVC drainage pipe and fittings. When making joints, the ends of the pipe should be cut squarely. A cleaner and primer should be used on the pipe before making a joint. When the pipe and fitting are properly prepared, swab glue on both of them and insert the pipe into the hub of the fitting. Once the pipe bottoms out, turn it about a half a turn. These joints can take a long time to set up. If you move the connection before the glue has dried, the joint will leak.

Words of Wisdom

I would like to close this chapter with some words of wisdom. As a seasoned master plumber, I have experienced many challenges in the field of plumbing. There is not room in this book to explain all that I've learned, but let me tell you a few things that may help you in your career.

Don't drop CPVC pipe on a concrete floor in cold weather. The pipe becomes brittle when it's cold, and if it is dropped, it can crack. The hairline cracks may go unnoticed until you test your installation. Finding cracked pipes shortly before the plumbing inspector is scheduled to arrive is no fun.

Wear safety glasses when you are soldering. I still carry a scar in the corner of my eye from an accident with solder splashing off a pipe.

Always pay attention to where the flame of your torch is. Carelessness with an open flame can result in disaster.

Be careful if you are soldering a pipe that has water in it. Not only will it be difficult to solder, it can be dangerous. A long time ago I learned, first hand, the power of steam. My helper and I had completed a rough-in on a remodeling job. We tested the pipes with water and found a leak at a stub out for a tub spout.

We cut the water off and drained the system. I watched as my helper heated the copper stub out to resolder the joint. Water had collected in the pipe and steam began to build up. All of a sudden, the stub shot out of its fitting and embedded itself in the drywall at the end of the tub. If one of us had been standing in front of that copper missile, it is hard to say what would have happened. A lot of pressure can develop when you are soldering and turning water to steam, so be careful.

Avoid letting your water pipe lay in the dirt. When dirt gets into the ends of your pipes it will ultimately come out through the plumbing fixtures. This dirt can cause problems in faucets and valves.

As you learn your trade, learn how to find economical ways to route your piping. Many plumbers waste a lot of pipe because of poor planning. They create extra work for themselves and make the cost of the job go up. Look before you leap, and you will learn to find ways to conserve pipe.

I would like to go on with my words of wisdom, but space is limited, and this is already a long chapter. Let's turn over to Chapter 19 and see what we can find out about gas pipe.

Chapter

19

Gas Piping

Gas piping is not a job that all plumbers do, but it is not unusual for plumbers to install gas systems. Gas piping is similar to running water pipe, but there are some significant differences. Not all local plumbing codes include sections on working with gas, but many do. The two types of gas plumbers normally work with are natural gas and liquefied petroleum (LP) gas.

Do you need to know anything about gas piping? That depends on the customs in your locality. If plumbers install gas pipe in your area, you should learn how to do it. Will your plumber's test contain questions on gas piping? Again, that depends on where you take the test. Not all jurisdictions consider gas piping a part of plumbing; therefore, the licensing exam for plumbers will not contain questions on the subject. Other areas do require plumbers to be knowledgeable of gas procedures. And then there are places, such as Maine, that do not require gas fitters to be licensed—you figure this one out; I can't.

Working with gas can be extremely dangerous. The fumes from gas can cause death, and the explosion possibilities are awesome. How any place could not regulate the installation and service of gas piping with licensing laws is beyond me. When you work with gas, you must pay attention to what you are doing; it is no time to be daydreaming.

The rules and regulations of gas piping are typically found in the mechanical code rather than in the plumbing code. However, it is not unusual for plumbing codes to provide information on the subject. This chapter is going to show you the basics of gas piping. It will not cover every aspect of the job or the code, but it will provide you with a foundation of knowledge.

I have worked with gas piping for going on 20 years, and I hold a master's license in both plumbing and gas fitting. The majority of my work experience has been in Virginia, with natural gas. I have, however, worked with LP gas since moving to Maine. I am about to share my understanding of good installation principles and my experience with you. Don't take this information as your ticket to being a gas fitter. Check with your local authorities for licensing information and code requirements before doing any gas work.

Materials for Gas Piping

Materials for gas piping vary, but steel pipe is the most common type of material used to convey gas to equipment and appliances. Copper tubing is often use to make short connections between the steel pipe and the gas appliance. Flexible connectors are also frequently used for this purpose. What types of materials are approved for fuel gas pipe? I'll show you.

The fittings used in a gas system must also be of an approved type, and they must be compatible with the pipe being used. If flexible connectors are used, they must be labeled as an approved device, and they may not be more than 6 feet in length.

The materials commonly approved for use as fuel gas pipe include aluminum-alloy pipe and tubing, brass pipe, copper or copper-alloy pipe—usually type-L or type-K—ductile iron pipe, plastic pipe and tubing, and steel pipe and tubing. Unless a distinction needs to be made, when I refer to pipe for the remainder of the chapter, it can be any of these materials.

There are restrictions on where and why some of these materials can be used. For example, aluminum-alloy materials are restricted to above-ground, interior use only. It is prohibited from use where it will come into contact with moisture, masonry, plaster, or insulation.

Ductile iron pipe is limited to underground use outside of buildings. It may not be used within the confines of a structure, and it may not be installed above ground.

Metallic materials are only to be used on systems where the fuel gas will not corrode the piping. If metal pipes are going to be exposed to corrosive conditions, the pipe must be protected in an approved manner.

Before any of gas pipe can be bent, it must be approved for bending. If approval is given for bending a pipe, the bend must be made with equipment that has been approved for that purpose. It is, of course, wrong to bend a pipe beyond its structural limitations.

Putting Gas Pipe Together

When it comes to putting gas pipe together, there are dozens of rules and regulations. The types of joints and connections made with gas materials will depend largely on the type of material being used. All joints and connections must be of a type that is approved for gas piping, and obviously, they must be gas-tight. The joints will be subjected to a pressure test before the installation is approved for use.

If your local code provides regulations for gas piping, you may notice that in the section dealing with joints and connections there are numerous cross-references to other sections or articles. The nitty-gritty details of making approved gas joints will be determined by your local code.

There are some aspects of gas joints that are generic enough to talk about here. For example, all pipe ends should be cut squarely. The pipe should be reamed and chamfered, and all burrs and obstructions should be removed. The end of the pipe should retain its full diameter, and pipe dope—or tape intended as a thread sealant—should be applied only to male threads.

Brazed joints should be clean and an approved flux and filler metal should be used as required.

Mechanical joints should be installed within the guidelines provided by their manufacturers.

Welded joints must be clean and made within the limits of approved procedures.

Heat-fusion joints must be clean and dry. They should be heated to the melt temperature, joined, and left alone until sufficient time, as provided by the manufacturer's instructions, has passed. This type of joint is used on polyolefin polyethelene (PE) and polybutylene (PB) pipe. Polyolefin cannot be joined with solvent cement.

Joints for polyvinyl chloride (PVC) pipe require the use of an approved primer and solvent. The pipe and fitting surfaces must be clean and dry.

If different types of materials are being joined together, the joint must be made with an approved adapter fitting. Dielectric fittings may be required when joining different types of metallic materials.

Supporting Gas Pipe

Finding the distances allowed for intervals between hangers supporting gas pipe is done in the same way as it is for plumbing pipes. By referring to the proper text or table in the appropriate code, you can identify approved intervals for supports. I can't guarantee what the requirements are in your area, but I will give you some common accepted practices for supporting gas pipe.

Rigid pipe with a 3/4-inch diameter or less should be supported at intervals not exceeding 10 feet. Rigid pipe with diameters of 1 inch or larger can go 12 feet between supports. Tubing that has a diameter of

1 1/4 inch or less should be supported every 6 feet. If the tubing diameter is 1 1/2 inch or larger, the spacing can be in 10-foot intervals.

Valves

All valves used in a gas system must be approved for that use. Their materials and construction must be compatible with the pipe being installed in the system. Where are valves required? Every gas meter must have a valve installed on the supply side of the meter. All buildings that contain gas pipe must be provided with a cut-off valve on the outside of the building. In addition, all gas outlets to equipment and appliances must be equipped with individual cut-offs. The cut-offs must be accessible and adjacent to the unit it is serving. Gas valves may not be installed in air plenums. The types of valves used most often are called gas cocks.

Underground Work

Underground work involving gas pipe has its own set of rules. Gas pipe installed below grade should be buried at least 18 inches deep. Plastic pipe installed underground must be kept at a safe distance from any heat source, and all underground piping must be installed to allow satisfactory maintenance and protection.

As an exception to the above, underground piping for outside appliances, like gas grills, may be buried at shallow depths of at least 8 inches. However, this type of installation must be approved and the pipe must be located in a way that protects it from physical damage.

As was mentioned earlier, pipes installed where corrosion may occur must be protected from corrosive action. If different types of metal pipe are joined together, underground dielectric fittings are required.

If plastic pipe is connected to fittings with no flexibility in them, the plastic pipe must be protected with a sleeve.

Just as with plumbing pipes, underground gas pipes must be protected from flood conditions.

Gas pipe that penetrates a foundation wall must be sleeved. The same procedures used for sleeving plumbing pipes will work for gas pipe.

Above-Ground Work Outside of Buildings

In addition to normal safe installation practices, above-ground work outside of buildings must be supported well and protected from any form of physical damage. Laying gas pipe on top of the ground is not normally approved. Possible flood conditions are another form of damage that these pipes must be protected against.

Installing Gas Pipe in Buildings

Installing gas pipe in buildings requires the use of approved practices and techniques. If the pipe will be placed in concrete, it should be installed so that it is covered in all directions by at least 1½ inches of concrete. Under these conditions the gas pipe should not be allowed to touch any metal, like rebar, and it should not be installed in concrete that contains quick-set additives or cinder aggregate.

Gas pipe that carries any type of gas, other than dry gas, should be graded downward to a drip leg (Fig. 19.1). The amount of grade should be at least 1/4 inch for every 15 feet of distance the pipe travels. All drip legs must be accessible, and they may not be installed where they will be subject to freezing temperatures.

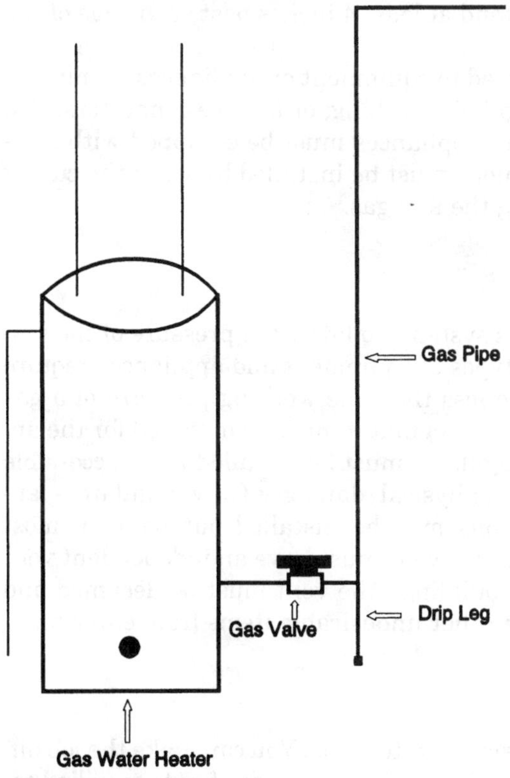

Figure 19.1 Gas drip leg.

When you take off from a main gas pipe with a branch pipe, the take-off must be made from the top or the side of the main. It is not permissible to drop from the bottom of a horizontal primary pipe. Just remember that your branch tee must turn up or out, never down.

Gas piping can be concealed, but when it is, it may not contain unions, tubing fittings, or running threads. It is not permissible to install gas piping in chimneys, vents, dumbwaiters, supply air ducts, laundry chutes, or elevator shafts.

If gas pipe, other than black or galvanized steel pipe, is installed concealed in a way that it may be hit by a nail or screw, kick plates must be installed to protect the pipe. If there are at least 1¼ inches of wood between the pipe and the potential source of puncture, kick plates are not required, but they are still a good idea. When kick plates are required, they must be made of steel with a thickness of at least 1/16 inch. The plate must be large enough to cover the area of the pipe where it passes though the notched or drilled area. Further, the kick plate must extend at least 4 inches above, below, and or beside the hole or notch. In other words, if the gas pipe is passing horizontally through a wall stud, the kick plate must extend at least 4 inches past each edge of the stud.

When gas piping is connected to equipment or appliances, it must be connected with approved rigid pipe, tubing, or flexible connectors. The outlets for all equipment and appliances must be equipped with individual cut-off valves, and unions must be installed between the cut-off valves and the unit receiving the fuel gas.

Regulators

Regulators are used with gas systems to lower the pressure of the system at an appliance. Many types of equipment and appliances require operating pressures that are less than the working pressure of a gas system. When this is the case, a regulator must be installed for the appliance or equipment. The regulator must be installed in an accessible location and protected from physical damage. Only regulators approved for outside applications may be installed outdoors. In most cases, a regulator that requires a vent must have an independent vent that terminates outside the building. The vent must be designed and installed to prevent water or other undesirable items from entering it.

Sizing Gas Pipe

Sizing gas pipe is similar to sizing water pipe. You can make the job difficult, or it can be reasonably simple. With the use of code-supplied tables and text, it is fairly easy to size simple gas jobs.

To size gas pipe by using tables, you must know the following: system pressure, pressure drop, specific gravity, and the total developed length of the pipe system. Once you have this information, the sizing tables will do the work for you.

The typical maximum operating pressure for a regular gas system is 5 pounds per square inch. There are exceptions to this rule. Some types of buildings and gas pipe are allowed higher pressures.

LP-gas systems are normally confined to a maximum operating pressure of 20 pounds per square inch gravity. There are also exceptions for these gas systems.

Dispensers

Gas dispensers must have emergency switches that will allow the power to the dispenser to be shut off. The gas supply pipe to a gas dispenser should have an approved backflow preventer installed on it.

Backup Gas

Backup gas systems are sometimes used. If standby gas is interconnected with a regular gas piping system, approved safety equipment, such as a three-way, two-port valve, must be installed. The prevention of backflow into either supply system must be avoided.

There may be times when an appliance requires supplemental gas to be provided in addition to regular gas. When this is the case, a backflow preventer must be installed on the gas pipe to the equipment requiring supplemental gas. The backflow preventer must be located between the source of the supplemental gas and the gas meter.

Testing

Gas pipe must be tested before the system can be put into use. Any part of the system that will be concealed must be tested prior to concealment. Most gas fitters test their work with air, but any number of gases can be used to conduct the test. Gases that are normally approved for testing gas systems include neon, helium, radon, nitrogen, carbon dioxide, xenon, nitrogen, argon, and krypton.

The required pressure for testing is normally equal to 1½ times the system's normal working pressure. However, the pressure test should never be conducted at a pressure of less than 3 pounds per square inch. Systems using low-pressure LP gas should be tested at the pressure of an 18-inch water column. All gas tests should be conducted for a duration of at least 10 minutes.

The test equipment used for gas systems can be a water manometer, mercury gauge, or other approved device. When a gas system is purged, the purging must be done in a safe location. The combustion chamber of gas equipment or appliances is not considered a safe location.

Looking for Leaks

Looking for leaks in gas pipe the wrong way can be deadly. A handful of plumbers look for leaks in natural-gas piping with an open flame, either from a torch or candle. The flame will find a gas leak, but you may not live to repair it. Don't check for gas leaks with an open flame, especially in LP-gas systems.

The most common safe way to find gas leaks is with the use of soapy water. Mix a solution of soap and water and apply the liquid to the threaded area of your pipes. If a joint between the pipe threads and a fitting is leaking, the water will bubble. You will have found your leak without risking an explosion.

Closing Comments

I have a few closing comments on the subject of working with gas. Personally, I feel anyone allowed to work with gas should be licensed. Since I don't make the rules, some places allow anyone to install and service gas systems. If your area does not require licensing for gas work, don't take the freedom lightly. Find, read, and understand everything you can about gas procedures. You are working with a substance that can kill, and kill quickly. Never underestimate the power of gas.

Chapter

20

Plumbing Fixtures

When we talk of plumbing fixtures, we are talking about many types of materials. A hose bibb is a plumbing fixture and so is a toilet. These two fixtures don't look anything alike, and the methods used to install them are as different as hot and cold, but they are both plumbing fixtures.

Apprentices generally get a lot of experience with plumbing fixtures. They get to load and unload them from trucks. Putting faucets on lavatories and sinks is usually a job handed over to apprentices. The assembling of toilets is frequently entrusted to apprentices. Experienced apprentices are allowed to install the connections for water heaters. In general, apprentices get a lot of hands-on experience when the work involves fixture installation.

This chapter is going to show you what the most common standard plumbing fixtures are. In addition, you will be told how to install these fixtures. There will even be advice on how to avoid breaking fragile fixtures; many apprentices break toilets when tightening the bolts used in the installation procedures. You will get some inside advice on how to make your installations go more quickly and more smoothly. All in all, you will learn a lot about plumbing fixtures.

This chapter covers all of the standard plumbing fixtures, including water closets, urinals, bidets, lavatories, kitchen sinks, garbage disposers, bathing units, water heaters, silcocks, hose bibbs, laundry hookup boxes, dishwashers, floor drains, faucets, laundry tubs, bar sinks, and fixtures for handicapped users.

Hose Bibbs and Silcocks

Let's get the ball rolling with some simple fixtures, like hose bibbs (Fig. 20.1) and silcocks (Fig. 20.2). First of all, every plumbing fixture installed must be of an approved type. Your local code book will describe the approval requirements which must be met by fixtures used in your locality.

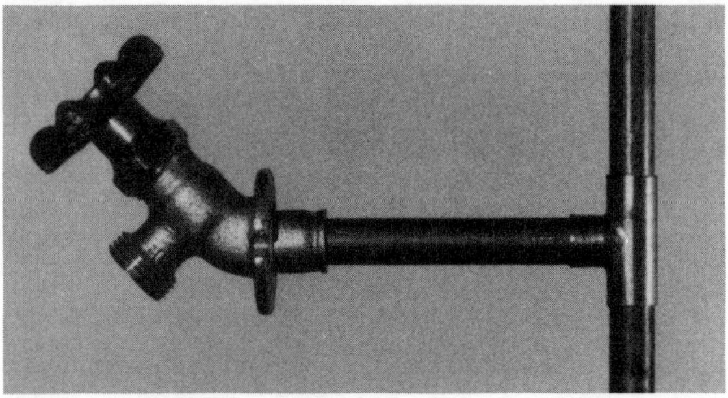

Figure 20.1 Hose bibb.

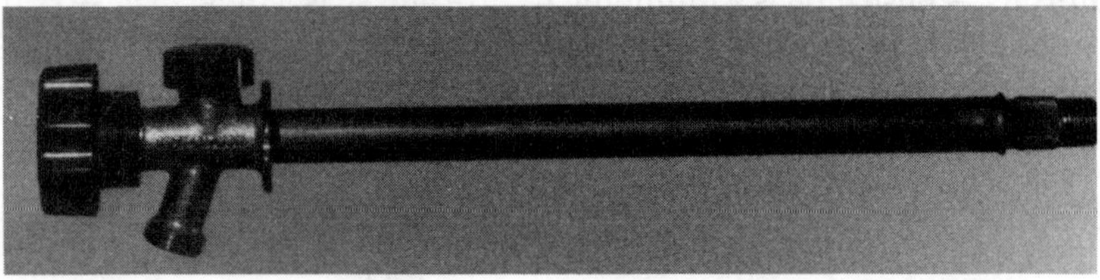

Figure 20.2 Frost-free hose bibb.

As you probably know, hose bibbs and silcocks are the devices mounted on the exterior walls of buildings for use as outside faucets. Both of these devices have threads to accept a standard garden-hose connection. Therefore, these fixtures must be protected against backflow, usually with a vacuum breaker. Some silcocks have backflow protection devices built into them. If the hose bibb or silcock being installed doesn't have an integral backflow preventer, you must install some type of backflow preventer.

The most common add-on type of backflow preventer for these fixtures is a small vacuum breaker that screws onto the hose threads. Once the vacuum breaker is in place, a set screw is tightened into the side of the vacuum breaker until it breaks off. When a garden hose is connected to the fixture, it is connected to threads on the vacuum breaker. Since the vacuum breaker is between the hose and the potable water supply, the water distribution system is protected against backflow.

Silcocks are often made to be what is considered frost-free. These fixtures mount on the outside wall of a building, but when they are turned off, there is little to no water left in the piping that occupies unheated space. Through the use of a long stem, the washer and seat of the silcock can be several inches from the spout of the fixture. When the valve is closed, water drains out of the piping in the unheated space, and the silcock is not supposed to be affected by freezing temperatures. They do sometimes freeze and burst, but this is usually due to poor installation procedures.

Silcocks are available in various lengths because of differences in the thickness of walls occupied by frost-free silcocks. Normally the inlet end of this type of silcock will extend through an interior wall and be exposed to heated space. Sometimes the back end of the silcock will be concealed behind a wall covering. In all cases, a cut-off valve must be installed on the supply pipe to the silcock, and the valve must be accessible.

When you install a frost-free silcock, you should install it so that any water remaining in the housing after the valve is closed will drain out. If the silcock is installed dead level or with a backward pitch, the effectiveness of the fixture is diminished.

To install silcocks, you must have a hole in an exterior wall. Once the hole is made, the silcock slides through it. There will be a mounting flange surrounding the silcock at the end where the spout and handle are located. These flanges are typically mounted to the exterior wall with the use of two screws. Any open air space behind the flange and around the pipe should be filled to prevent air infiltration. Mortar mix, insulation, or expanding foam is normally used to fill these gaps.

The connection end of a silcock will typically accept a solder or screw connection; the choice is yours. I normally solder the connection, but some plumbers prefer to use female adapters in making screw connections.

Installing a hose bibb is very similar to installing a frost-free silcock. However, hose bibbs do not have the extended stem housing that frost-free silcocks do. Therefore, you must install a section of pipe to reach through the exterior wall, between the hose bibb and the interior cut-off valve. Typically, this section of pipe is soldered into the hose bibb.

Floor Drains

Floor drains are plumbing fixtures, but they are normally installed during the rough-in. These fixtures are installed with the same basic techniques used to install drainage pipes. Floor drains are required to have removable strainers so that their traps are accessible. The inlet to the drain, however, must be readily accessible. What is the difference between accessible and readily accessible? Accessible means that something can be accessed with minimal trouble. For example, if a door has to be opened to gain access to an object, the object is accessible. Readily accessible means you must be able to lay your hands on an object without overcoming any obstruction, such as a door. Most codes require that all floor drains have a minimum diameter of 2 inches.

Laundry Boxes

Laundry boxes are often used to house the water and drain pipes for automatic clothes washers. These boxes are generally installed during the rough-in phase. The boxes have nailing flanges that allow them to be attached to wall studs, and the boxes are normally recessed in a wall.

The drain connection is basically a coupling for the drainage pipe to connect to. Some boxes require threaded adapters for making the drain-to-box connection. The drainage hose from the clothes washer simply sticks into the standpipe. These standpipes normally have a 2-inch diameter and range in height from 18 to 30 inches.

The water connections consist of two boiler drains. The boiler drains sit in the box and are connected to the water distribution system either above or below the box. Since boiler drains have hose threads, you may be required to equip each of them with a screw-on vacuum breaker. Boiler drains normally offer the option of soldered or screw joints for connection to the water distribution pipes.

Bathing Units

Bathing units break down into two basic categories: bathtubs and showers. These fixtures are installed a little differently. Therefore, we will look at each type of installation, individually.

Bathtubs

Bathtubs in most new houses combine a bathtub with an overhead shower. Some of these units are formed as one-piece units. Others consist of only a bathtub. If a shower facility is added, the surrounding tub walls are waterproofed with tile or some other approved tub-surround material. Installation procedures for plain bathtubs are not the same as the techniques used for one-piece tub-shower combinations. Let's start this lesson with standard bathtubs.

Standard bathtubs. Standard bathtubs might be made of heavy cast iron, lightweight plastic, or a number of other approved materials. A standard bathtub is 5 feet long. To make installation easy, the rough opening for the tub must be slightly larger than the fixture. These openings are usually made up to 1/2-inch larger than the tub. In all installations, for any plumbing fixture, you should follow the manufacturer's recommended installation procedures.

Standard bathtubs need support where they meet the back wall of their opening. The front of the tub is supported by its skirt or apron. There are two common ways of supporting the edge of the tub at the back wall. You can use a strip ledger or vertical supports. These supports must be installed at a height that will make the tub level.

You can either read the installation instructions to see how high above the floor the supports should be, or you can make your own measurements. Set the tub in open floor space and support it by hand until it is level. Then measure from the bottom lip of the area you will be supporting to the top of the floor. This will give you the proper height measurement for your supports.

If you are going to use a strip ledger, cut a piece of wood to a length just under 5 feet. The wood can be a 2- by 4-inch wall stud. Place it against the bare studs of the back wall, position it at the proper height, and nail it to the studs. The ledger should be installed horizontally. Check the ledger strip to be sure it is level, and then you are ready to put the tub into place.

If you are going to use vertical supports, cut your lumber to the proper length. You will need two supports, one for the front, or head, of the tub and one for the back. Place the wood vertically at each end of the back wall and nail it to the wall studs. Lay a long level across the span of the supports and check to be sure the tub will set level when placed on the columns.

When your supports are installed and checked for level, you can position the tub in the opening. This can be a little tricky with a 400-pound cast-iron bathtub. Don't get your fingers under the edge of the tub where they will be mashed or severed. When you slide the tub into the opening, it must be positioned properly. If it is going into the hole at an angle, it won't fit.

Slide the tub into the hole and lift it up and onto the supports. You may have to sit on the floor and use your feet and legs to slide a heavy tub into its opening. Once the edge of the tub is touching the exposed studs of the back wall and sitting on its support, check to see if it is level. If it is, you can move on to the next step.

You can either install the tub faucet or the tub waste next. Let's move ahead with the tub waste and overflow. Tub wastes come in a variety of forms and are made out of various materials. If the tub waste and overflow you use will be held together with slip nuts, access must be provided to the fittings. If you use a plastic tub waste where the joints are glued together, no access door is required. Most people don't appreciate access doors in their finished walls, so plumbers often install tub wastes that don't rely on mechanical joints. However, if you are installing a tub waste with mechanical joints, Figs. 20.3 through 20.17 show you how to make the installation.

Let me say now that there are so many different brands and types of plumbing fixtures, equipment, and trim that an entire book could be dedicated to just those subjects. Since space is limited in this book, I will be picking specific examples to explain to you. Don't assume what you are reading is the only way to get the job done. It is one way, but there are others.

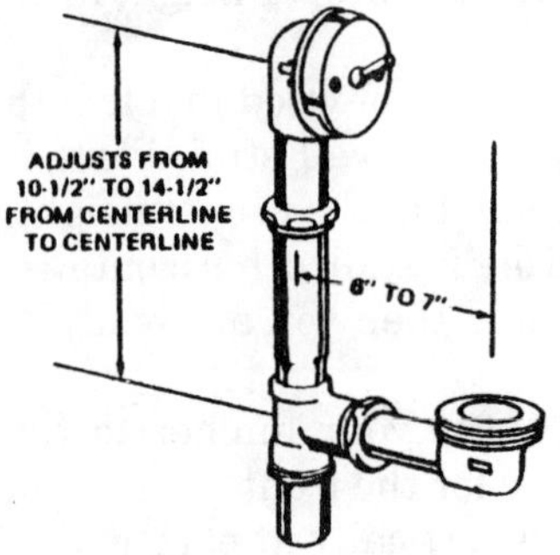

Figure 20.3 Installing a tub waste. (*Moen, Inc.*)

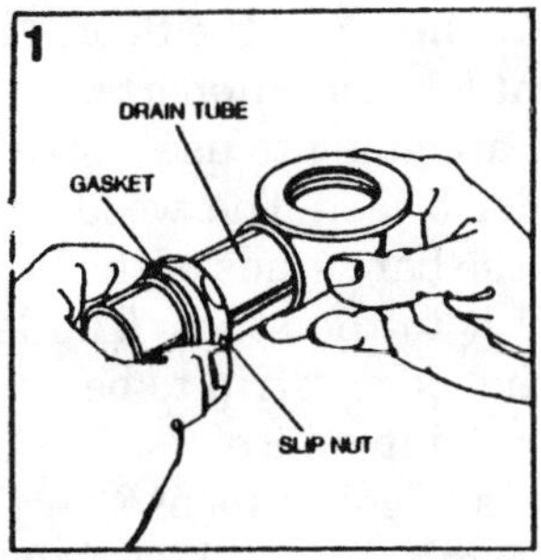

1. Place slip nut and gasket on drain tube.

Figure 20.4 Installing a tub waste. (*Moen, Inc.*)

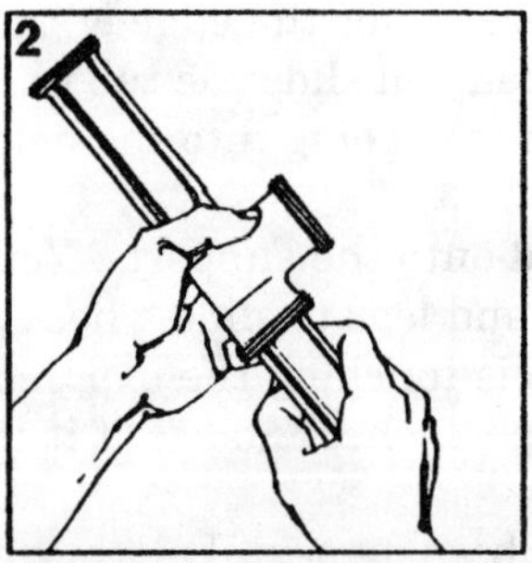

2. Screw the tailpiece up into the bottom of the drain tee. Use pipe dope on threads. Make tight.

Figure 20.5 Installing a tub waste. (*Moen, Inc.*)

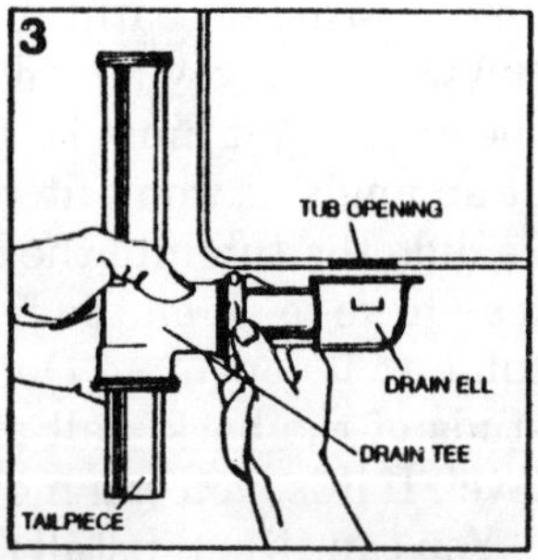

3. Place drain tube into the drain tee. Hold the drain against the bottom of the tub at the opening and tighten slip nut hand tight.

Figure 20.6 Installing a tub waste. (*Moen, Inc.*)

4. This positions the waste assembly so that the connection to the waste pipe opening can be determined, whether by slip nut, direct connection to a 1½" waste pipe thread, or any other method of your own choosing.

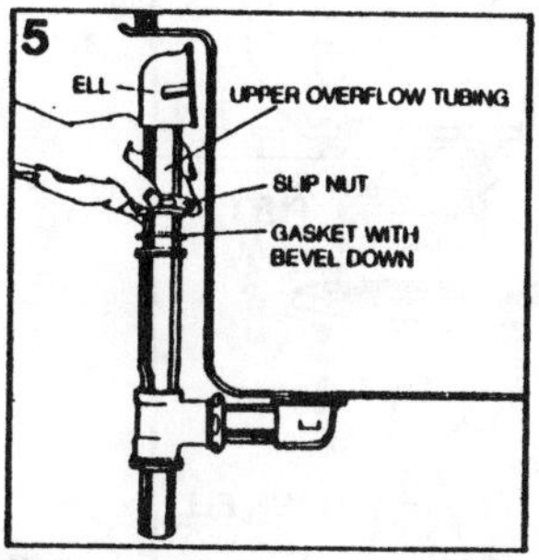

5. Place the slip nut and gasket on the overflow tube. The bevel on the gasket

Figure 20.7 Installing a tub waste. (*Moen, Inc.*)

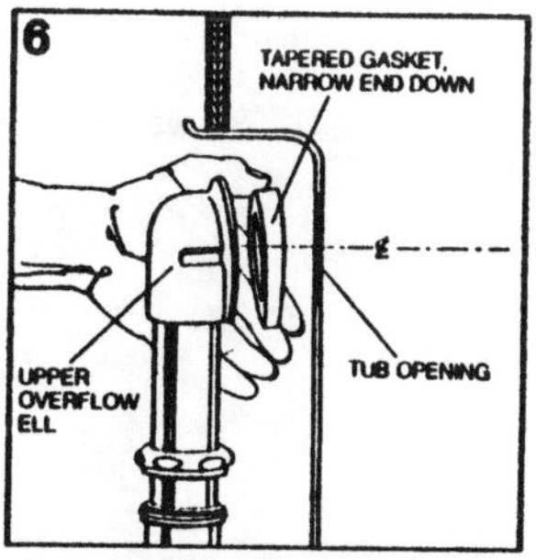

6. Center the upper overflow ell with the tub opening. Place the tapered gasket between the flange on the ell and the tub opening, – narrow side down.

Figure 20.8 Installing a tub waste. (*Moen, Inc.*)

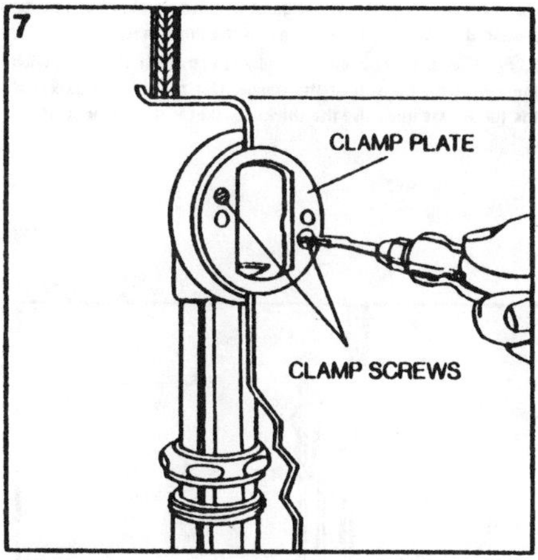

7. Center the clamp plate over the tub opening and thread the brass clamp screws through it into the overflow ell. Use the 4 o'clock and 10 o'clock positions. Tighten screws making sure all parts are aligned.

Figure 20.9 Installing a tub waste. (*Moen, Inc.*)

For this example, we will assume that you are using a plastic tub waste with a pop-up stopper. These are the easiest to install and the require minimal maintenance. The pipe and fittings will be schedule–40 plastic, either ABS or PVC.

Take the tub shoe—this is the piece with female threads in it—and place the gasket that was supplied with the tub waste on the top of the flat surface. Then apply plumber's putty around the bottom rim of the

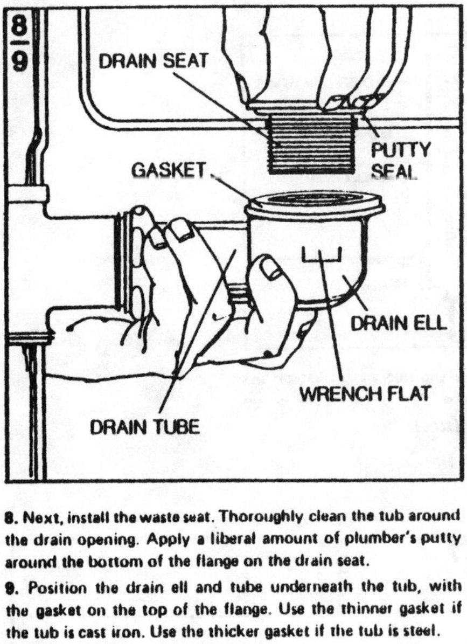

Figure 20.10 Installing a tub waste. (*Moen, Inc.*)

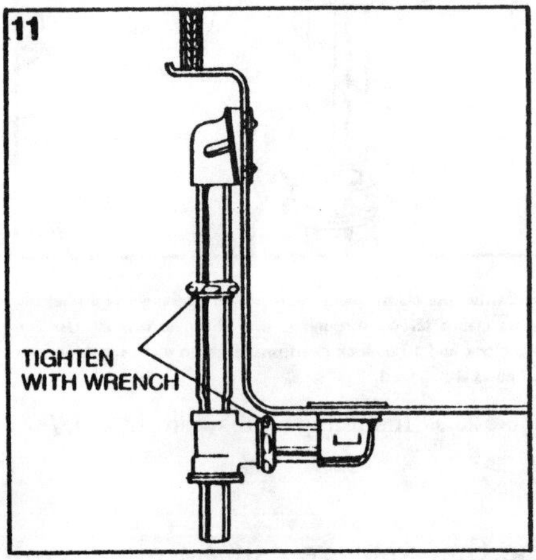

Figure 20.11 Installing a tub waste. (*Moen, Inc.*)

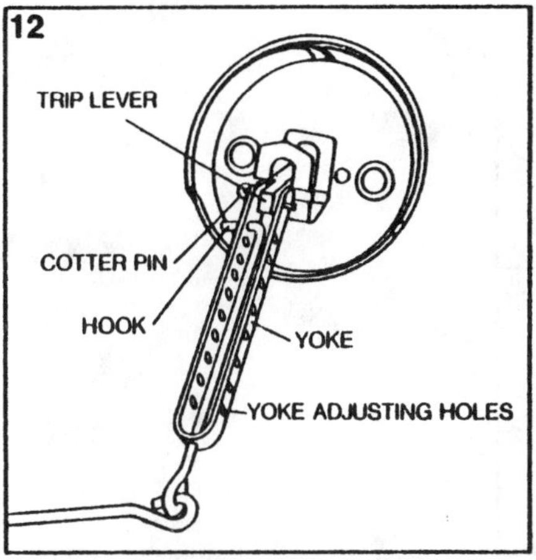

12. Remove the cotter pin in the top of the yoke in the lift assembly. Assemble the lift lever to the trip lever using the cotter pin.

Figure 20.12 Installing a tub waste. (*Moen, Inc.*)

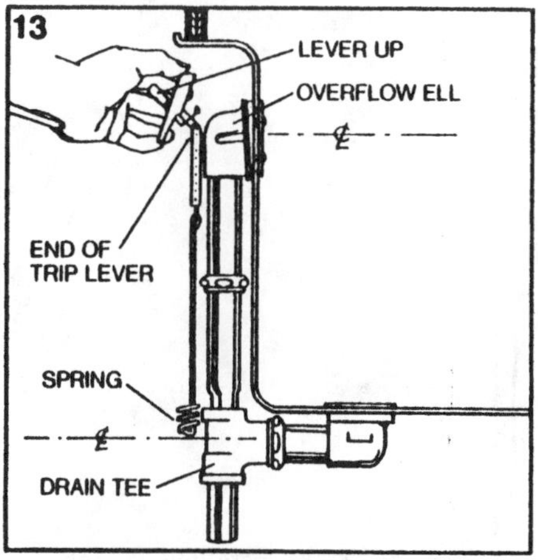

13. Position the trip lever UP, the lift assembly DOWN, and hold the lift assembly against the overflow tubes. Hold the end of the trip lever in the center of the overflow ell. The end of the spring should be in the center of the drain tube and tee. If it is not, adjust the lift length by moving the hook in the yoke adjusting holes.

Figure 20.13 Installing a tub waste. (*Moen, Inc.*)

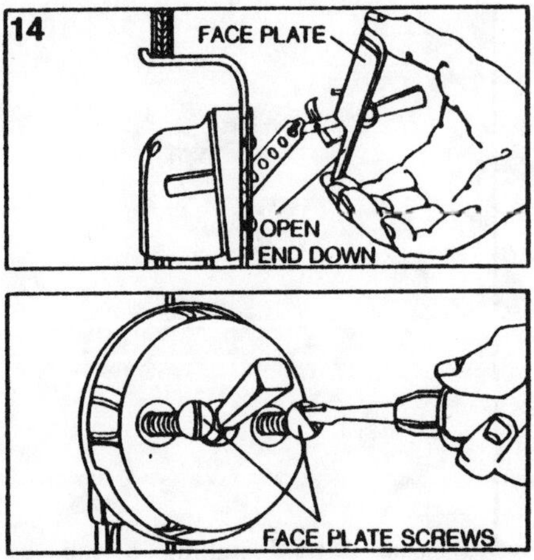

14. With the open side of the face plate DOWN, push the lift assembly down into the overflow tube. Thread the chrome plated face plate screws into the overflow ell, 9 o'clock and 3 o'clock.

Figure 20.14 Installing a tub waste. (*Moen, Inc.*)

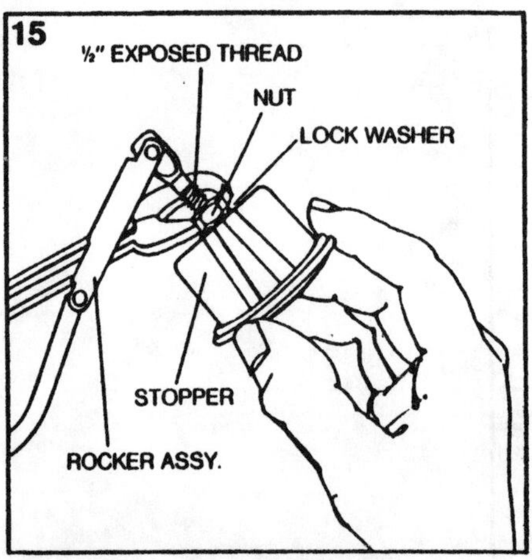

15. Remove the brass nut and star lockwasher from the end of the rocker assembly. Replace, putting on nut first and lockwasher last. Leave about ½" of thread exposed on the end. Thread on stopper and tighten the nut against it.

Figure 20.15 Installing a tub waste. (*Moen, Inc.*)

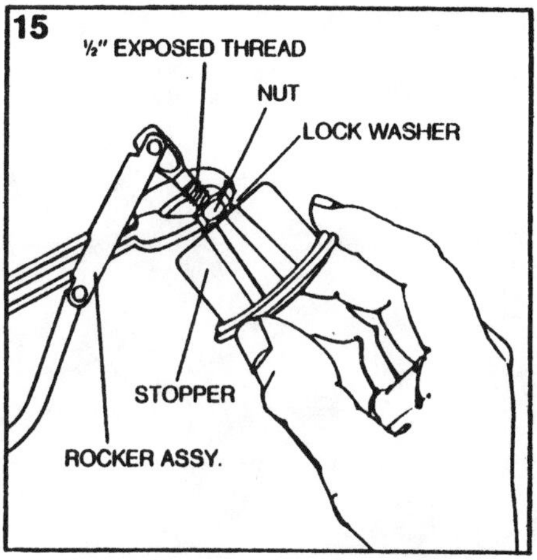

15. Remove the brass nut and star lockwasher from the end of the rocker assembly. Replace, putting on nut first and lockwasher last. Leave about ½" of thread exposed on the end. Thread on stopper and tighten the nut against it.

Figure 20.16 Installing a tub waste. (*Moen, Inc.*)

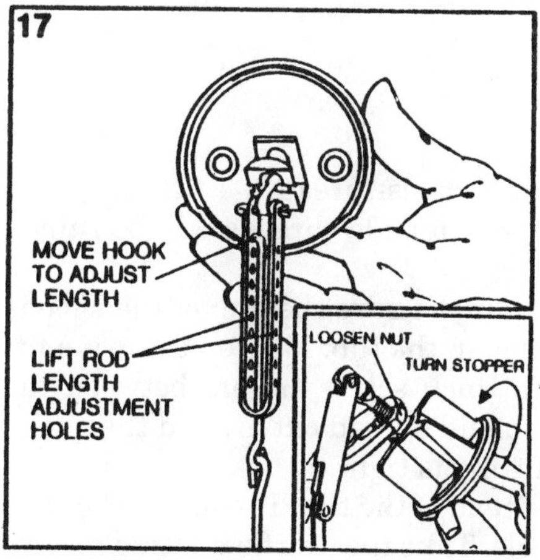

17. Adjust lift rod length until the lever opens and closes freely. Adjust stopper screw. If stopper raises to far and won't drop down, loosen the nut and screw the stopper down a couple of turns at a time until it drops freely.

18. Fill the tub. Drain and check all joints.

Figure 20.17 Installing a tub waste. (*Moen, Inc.*)

finished trim that has male threads (Fig. 20.18). Next, apply pipe dope to the threads.

You must position the tub shoe, with the gasket in place, under the tub so that it lines up with the drain hole. Holding it in place—this job works better with two sets of hands—screw the trim piece into the shoe

Figure 20.18 Applying plumber's putty to tub drain.

from inside the tub. Be careful not to cross-thread the fitting. If you cross-thread it—install it crooked—the female threads will be ruined and the joint will leak.

When the fitting is snug, check to make sure the hub of the shoe is pointed back toward the overflow hole in the tub. Tighten the drain fitting into the shoe by crossing two thick screw drivers between the cross-bars of the drain. The putty should spread out around the drain, on the inside of the tub when the fitting is tight.

Now, move to the back side of the head of the tub. The hub of the shoe should be pointed at the overflow hole. Take the overflow fitting and install the sponge or rubber gasket that came with it. The gasket is tapered; the thick part should be positioned at the top. From the front of the tub, attach the trim for the overflow hole. It will be held on with one or two screws.

Take the sanitary tee that came with the package and hold it in line with the hub of the shoe fitting. Make the necessary measurements and cut two pieces of pipe. One will connect the tee to the shoe, and the other will connect the top of the tee to the overflow fitting. The pipe coming out the bottom of the tee will connect to the bathtub's trap.

If a brass tub waste is used, a trap adapter will be required for the trap connection. A trap adapter accepts a glue joint with plastic pipe on one end and a slip-joint connection on the other end.

Now we will install the faucet, or valve as it is normally called. In this example we are using a single-handle, tub-only valve. The hot and cold water supplies will be connected to the valve. The hot water pipe should be piped to the left side, as you look at the valve from inside the bathtub. In the center of the valve there will be an opening on the bottom. This is for the water supply to the tub spout.

Some tub spouts are held in place with set screws. They slide over a piece of copper and the screw is tightened. Other types screw onto male adapters. Look at your tub valve and pipe it accordingly. Again, it is important that you follow the manufacturer's recommendations. I just don't have the space to provide you with all the options and scenarios you might encounter. Remember to keep the tub spout the distance above the flood-level rim that is required by your local code.

After the finished walls are installed around the tub, attach the tub spout and valve trim. All of the major work for installing bathing units is done during the rough-in phase of a job. When you go back to the job to set fixtures and trim out the bathing units, all you will be doing for a tub is installing the valve trim and tub spout.

When you are trying to decide where to rough-in your tub valve and spout, refer to a rough-in book. Rough-in books provide you with recommended measurements for all plumbing fixtures. Every plumbing manufacturer should be able to provide you with a rough-in book for their fixtures. I will give you the rule-of-thumb measurements, but always check the manufacturer's recommendations.

Tub spouts are generally installed 4 inches above the flood-level rim of a tub. The faucet, or valve, is usually set 12 inches above the rim. Most tub wastes are located about 14 inches out from the back tub wall. The hole for installing a tub waste is usually about 4 inches wide and about 1 foot long. The hole starts at the head wall of the tub and runs toward the back end of the tub. The center of the 4-inch width should be between 14 and 15 inches from the back wall of the tub opening. The rough-in for the tub and shower valve depths depend on the type of valves being used. Refer to rough-in books or drawings to determine the proper exposure of the valve.

One-piece tub-shower combinations. One-piece tub-shower combinations are installed a little differently. These units have their surrounding walls already attached. They slide into the tub opening and rest against the stud walls. Once the units are in place and level, both horizontally and vertically, they are nailed or screwed into place. The nails or screws are installed through a nailing flange that is formed around the outside edges of the bathing unit (Fig. 20.19).

Figure 20.19 Nailing flange on a one-piece tub-shower combination.

Once these units are set in place, most of the rest of the installation is the same as described above. No ledger strips or vertical supports are needed on these units. Since tub-shower combinations need a shower head, you will have that additional work and a slightly different valve.

The four-port valve (Fig. 20.20) will have a hole in the top of it for the pipe rising to the shower head. The shower head outlet is normally roughed-in about 78 inches above the floor. A drop-eared, or wing, ell (Fig. 20.21) is used at the end of the shower head riser. This fitting has two flanges, or ears, that allow it to be screwed to a piece of wood backing that you will install between two wall studs. This will keep the shower head from rattling and being pulled out of the wall.

The drop-eared ell has a normal pipe socket on one end and female threads on the other. Your shower head will screw into these threads. During rough-in, a pipe nipple is screwed into the threads. This keeps the shower arm from being damaged during construction and makes it easier for drywall hangers to work around the protrusion.

When you trim out these units, you must install the shower arm and head. The shower arm will have a crook in it. The longer section of the arm is what screws into the wing ell. Slide the escutcheon that is packed with the arm over the long end so that the finished side of the escutcheon will be facing the tub. Dope up the threads and install the arm. Turn it until it is hand tight. Apply pipe dope to the threads on the other end and screw the shower head into place.

Figure 20.20 Four-port tub-shower valve.

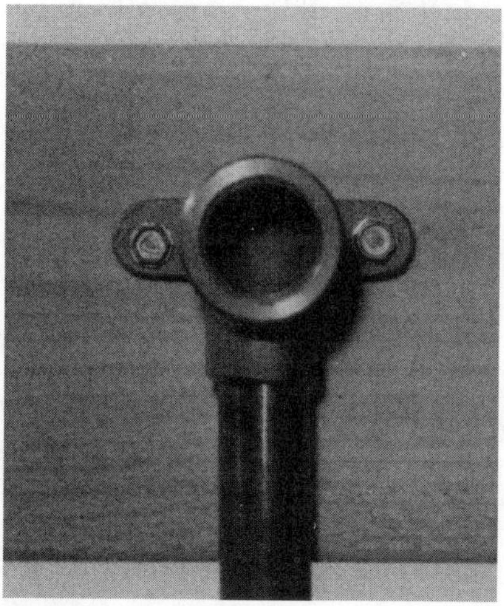

Figure 20.21 Drop-eared ell.

The shower head will have a collar with flat spots on it. This provides a place for a smooth wrench to be used in the final tightening. As you tighten the shower head, the shower arm will also turn and become tight. Do not use pliers or pipe wrenches on the shower arm; their teeth will cut into the finish and ruin the shower arm's appearance.

The height of the side walls that surround a bathtub that has an overhead shower should measure at least 6 feet above the finished floor level and not less than 70 inches when measured from the fixture's drain height. These walls must be waterproof and nonabsorbent.

Showers

The rules and procedures for showers are a little different from those for bathtubs. Let me show you what to expect.

One-piece showers. One-piece showers get set in place and secured just like one-piece tub-shower combinations. But remember to cut a hole in the floor for your drain before you install the bathing unit in its opening. These showers are installed during the rough-in work. Once the unit is in place, the shower valve should be installed about 4 feet from the finished floor. The shower head will rough-in at about 78 inches.

If the shower doesn't have an integral drain, you will have to install a drain. The shower drain is installed about the same way as a tub shoe. These independent drains are often made to accept the end of a

2-inch pipe. If they are, all you have to do is glue a piece of pipe between the drain and the trap and you are done.

Molded shower pans. Molded shower pans, or bases as they are often called, also get installed during the rough-in. A hole is cut for the drain and the pan is set into place. It is leveled and secured. You can't drive nails through the flange of most molded shower bases. However, driving roofing nails into the surrounding studs so that the big heads of the nails come over the wall flange will work to keep the base secure.

The drain collars on molded bases will not allow you to simply glue your drain pipe into place. You will either have to caulk the pipe into the drain with hot lead or use a rubber adapter. The adapters can be troublesome to get over the pipe, but they are what most plumbers use. Any plumbing supplier will know what you are talking about if you say you want a rubber adapter for a molded shower drain.

Poured shower bases. Poured shower bases have become somewhat rare. These are bases made from poured concrete and covered with tile. In the old days, the liners for these bases were made from lead or copper. Today, lightweight rubber-like material is used.

Poured shower bases are first framed up by carpenters. Then the plumber installs a special drain. The drain is adjustable in its height and resembles a floor drain. The bottom half of the drain is installed on the drainage pipe below the floor of the proposed shower stall. Then the pan liner is installed. A hole is cut in the liner, over the drain opening, and the top half of the shower drain is attached to the bottom half. The two halves are normally held together by several machine bolts. The strainer on the drain can be screwed in or out for a wide range of height adjustment.

The pan liner must cover the floor of the stall and must also turn up and extend up the walls, usually for at least 2 inches in height. The liner must also cover the threshold of the stall.

Most codes require shower stalls to contain a minimum of 900 square inches. A 30-inch disk should fit in the base of the stall. Some prefabricated showers don't meet this requirement. It would seem that a 30- by 30-inch, or 30-inch, shower as they are called, would meet the criteria. However the code is referring to interior measurement, and the shower size for a 30-inch shower is based on outside dimensions. Some code officers will issue variances that allow the use of 30-inch showers, but don't bet on it. If space limits the size of the shower stall you can use, get prior permission before using a stall that doesn't meet code requirements.

The height of side walls surrounding a shower should measure at least 6 feet above the finished floor level and not less than 70 inches when measured from the fixture's drain height. These walls must be waterproof and nonabsorbent.

Water Closets

Water closets are not hard to install, but they are also not hard to break. Most water closets are made of china, and china breaks very easily. To avoid embarrassment and financial pain, be careful when working with toilets. Don't drop them and don't get carried away when tightening their bolts.

The two most common styles of toilets are floor-mounted. One type has an elongated bowl (Fig. 20.22) and the other type has a round bowl (Fig. 20.23). Round-front toilets are normally used in homes and elongated ones are generally required if the public will have access to them. The seats for public and employee-use toilets are required to have a split front.

Water closets should be mounted to an approved floor flange. If the flange ring is made of plastic, it should have a thickness of 1/4 inch (Fig. 20.24). Metal flange rings can be 1/8 inch thick. The flanges should be mounted securely to the floor, typically with screws.

Toilets may be equipped with flush tanks, like the toilets in most homes, or flush valves, like the toilets in most public restrooms. Every toilet must have its own flushing device. There are code regulations dictating when and how use of each type of flushing device is determined. The code language on these issues is easy to understand, so I won't take the time or space to repeat it. A quick glance at your local code will explain when to use which type of device.

If you get into a lot of commercial work, you will also find that there are wall-hung toilets (Fig. 20.25). These toilets are installed much differently than floor-mounted toilets. We are, however, going to begin our installation instructions with a floor-mounted flush-tank toilet. The type we are going to install in this chapter is a two-piece toilet combination, like the toilets in most homes. There are also one-piece toilets (Fig. 20.26), but those are expensive and not used often in average homes or businesses.

Okay, we have our drain pipe roughed in and our closet flange installed. We are using a standard 12-inch rough toilet (Fig. 20.27). You can get 10 and 14-inch rough toilets if you need them, but 12-inch rough toilets are the standard (Fig. 20.28).

The distance from the back wall to the center of the drain makes the difference between these three toilets. Standard toilets use a drain with a center that is 12 inches from the back wall. Ten-inch rough toi-

Figure 20.22 Elongated toilet. (*Crane Plumbing Fixtures*)

Figure 20.23 Round-front toilet. (*Crane Plumbing Fixtures*)

Figure 20.24 Plastic closet flange with wax ring and closet bolts.

lets are roughed-in 10 inches from the back wall, and 14-inch rough toilets are roughed-in 14 inches from the back wall. These measurements are taken from the finished wall, so when you are roughing your pipes in, allow for the thickness of the finished wallcoverings.

While we are on the subject of rough-in dimensions, let me add that the center of your water closet's drain should not be closer than 15 inches to the closest object, on either side. In other words, you need a clear space with a width of 30 inches to install a toilet to code.

Now, you've got the tank and the bowl of the toilet out of the boxes. The first thing you do is set the closet bolts in the slots of the closet flange. They will move within the slot to allow proper alignment. Next, put a wax ring on the flange. Now you are ready to set the toilet bowl in place.

Set the bowl on the wax ring so that the closet bolts come up through

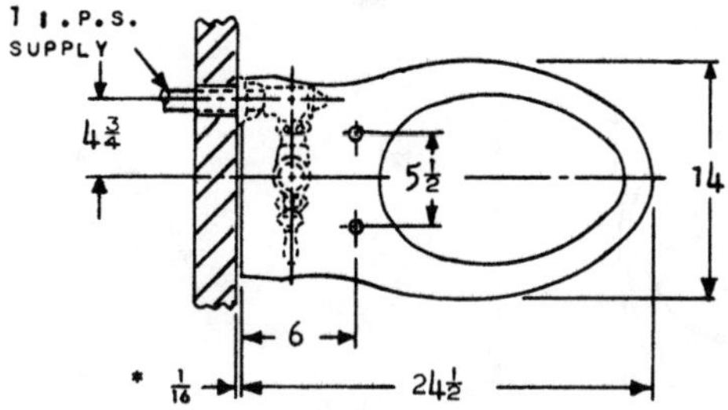

*These dimensions must be maintained for proper installation.

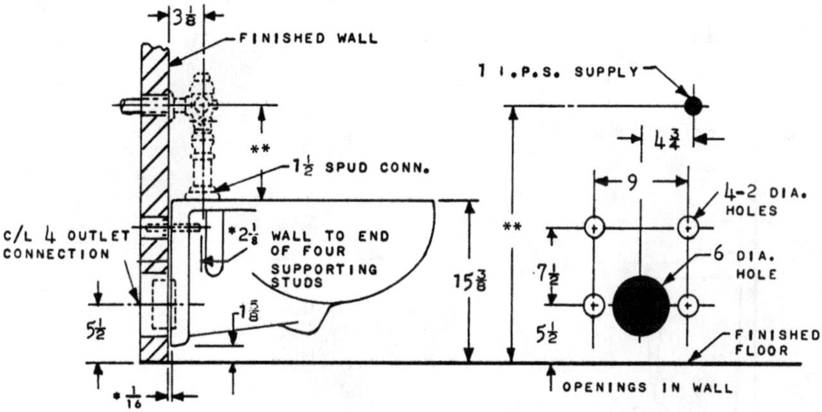

NOTE: Roughing-in dimensions for closet supporting unit to be obtained direct from manufacturer.
Check with manufacturer for proper outlet coupling extension beyond finished wall.

**Variable-check with manufacturer of flush valve for proper roughing-in dimension.

Figure 20.25 Wall-hung toilet. (*Crane Plumbing Fixtures*)

Figure 20.26 One-piece toilet. (*Crane Plumbing Fixtures*)

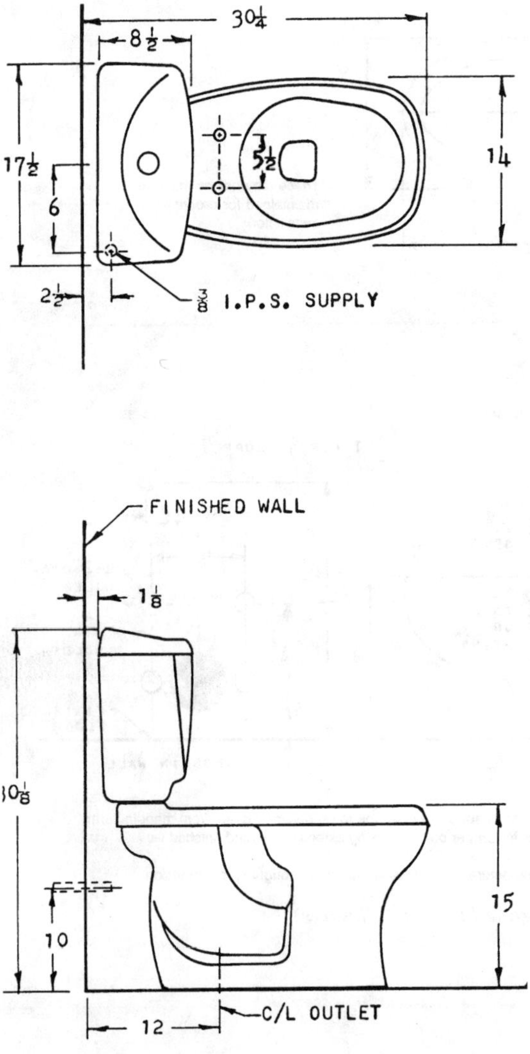

Figure 20.27 Standard 12-inch rough toilet. (*Crane Plumbing Fixtures*)

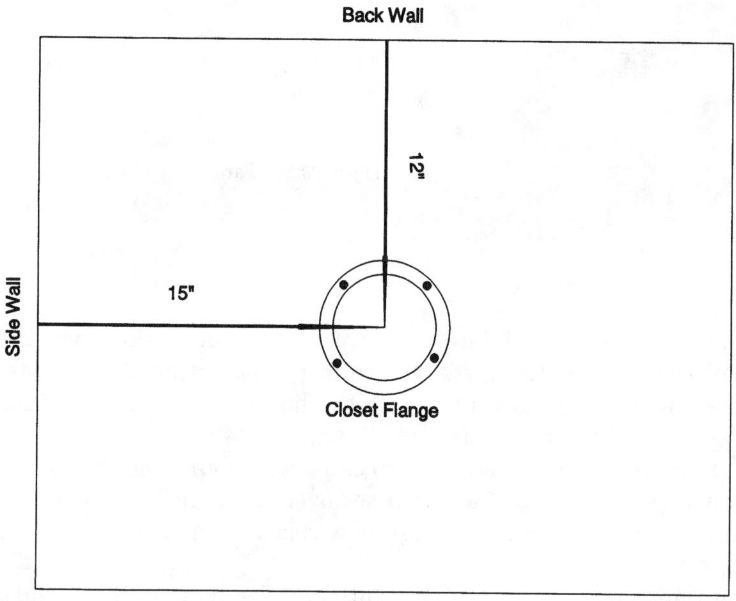

Figure 20.28 Standard toilet rough-in.

the holes in the base of the toilet. Press down hard. You may have to straddle the bowl and sit on it to compress the wax. Cold wax doesn't compress or seal well. If your wax ring is cold, warm it up before installing it. If you lay it on the dash of your truck on the way to the job, the defroster will warm it nicely.

With the bowl set in place, put the metal washers over the closet bolts. Then put the plastic disks that are packed with the toilet bowl over the bolts. Screw the nuts that came with the closet bolts down on the threads until they are hand tight. Use a wrench to snug the nuts up, but don't make them too tight. Too much stress will crack the toilet bowl.

Before you finish tightening the closet bolts, make sure the toilet is straight. You can measure from the back wall to the holes in the toilet where the tank connects to confirm its position. Tighten the bolts only until the bowl will not shift from side to side. Cut the bolts off slightly above the nuts and snap the plastic covers into place on the plastic discs.

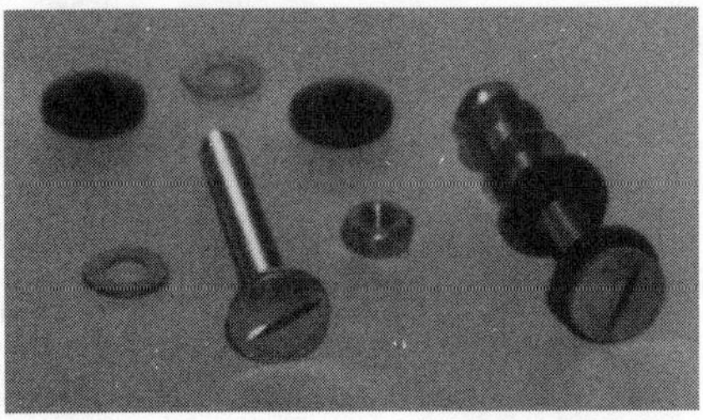

Figure 20.29 Tank-to-bowl bolts.

Lay the toilet tank on the floor, and press the big gasket over the threads of the flush valve. Slide the rubber washers on the tank-to-bowl bolts and stick the bolts through the holes in the bottom of the tank. The heads of the bolts should be inside the tank.

If the tank-to-bowl bolts (Fig. 20.29) came with four washers—some only come with two—slide the extra set over the threads protruding from the bottom of the tank. You are now ready to mount the tank to the bowl.

Set the tank in place so that the bolts and the bottom of the flush valve line up with the holes in the bowl. Hold onto the tank; if it falls off the bowl, it will probably break. Install the metal washers on the bolts and secure them with the provided nuts. Snug the nuts up with a wrench, but don't become overzealous. Again, too much tension will pop the china. By the way, it's a good idea to wear eye protection when working with china fixtures.

When the tank is seated firmly on the bowl, you are ready to connect the water supply. With the main water supply cut off, prepare the stub out for a stop valve. If the stub out comes up through the floor, you will use a straight stop (Fig. 20.30). When the fixture supply comes out of the wall, an angle stop (Fig. 20.31) will be used. I prefer compression stops, but some plumbers like the kind of stops that are soldered onto the supply, and others prefer to use threaded chrome nipples and stops.

Before you install the stop valve, slide an escutcheon over the supply pipe. Then install the cut-off. Remove the nut from the ballcock at the bottom left and front corner of the tank. Take a closet supply tube (Fig. 20.32) and measure it for the proper length to connect the stop valve to the ballcock threads where you removed the nut at the bottom of the tank.

Cut the supply tube with roller cutters and put the small end into the supply stop, but don't tighten the compression nut yet. Check your fit. If the supply tube fits, remove it from the stop and slide the ballcock nut over it, with the threads pointing upward.

Screw the ballcock nut onto the threads; pipe dope is not needed. The

Figure 20.30 Straight stop.

Figure 20.31 Angle stop.

Figure 20.32 Closet supply.

flanged head of the closet supply will seal the joint. Then, with the compression nut and sleeve on the bottom of the supply tube, insert it into the stop valve and tighten it.

Installing the seat is all that is left. Take the seat from its box and install it so that the holes in the hinge flanges line up with the holes in the toilet bowl. Apply the nuts to the seat bolts and you are done. An experienced plumber can complete this job, from start to finish, in about 15 minutes.

Toilets equipped with flush valves (Fig. 20.33) go together a little differently. The bowl will mount in the same way, but there is no tank to install. Instead, you must install the flush valve. Basically, the bottom of the flush valve mounts into the top of the toilet bowl. It is secured with a spud connection. The flushing mechanism is supported by the tube mounted in the bowl. Then a pipe extends out of the flushing mechanism and connects to a water supply coming out of the wall with a threaded connection. The pipes are concealed with chrome sleeves to give a better appearance.

Flush valves are equipped with flow controls and can only be used in systems where the water pressure is adequate for them to work properly. There are different types of flush valves available, so consult the manufacturer's recommendations for specific installation instructions.

Wall-hung toilets are not common. When these toilets are used, a carrier that is mounted in the wall during rough-in provides the support for the toilet. The toilet is mounted to bolts extending from the carrier and held in place with nuts. There will be a gasket used at the

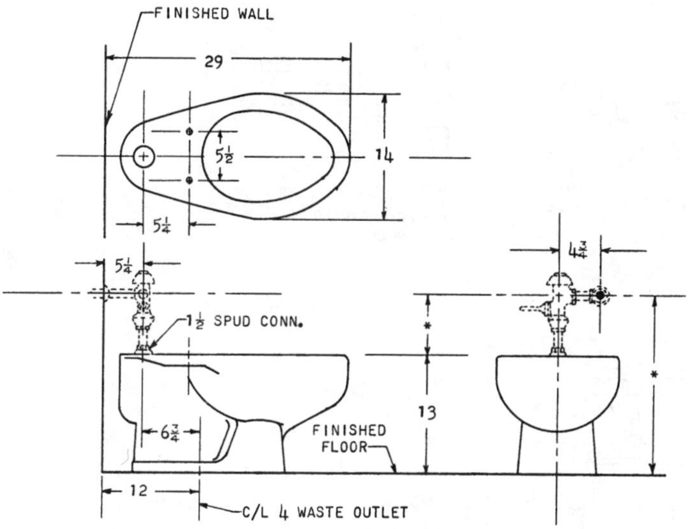

Figure 20.33 Flush-valve type toilet. (*Crane Plumbing Fixtures*)

point of connection to prevent leakage. Again, there are many types of wall-hung toilets and carriers. Consult the manufacturer's instructions before attempting installation of these toilets.

Urinals

I have never seen a urinal in a residential home, but there are plenty of them in commercial buildings. Urinals typically hang on a wall and are cleansed with the use of a flush valve. When designing a plumbing system, urinals can generally be used as a substitute for up to half of the toilets required. When installed for public use, urinals are required to have a visible water trap seal, without strainers. The walls and floor in the immediate vicinity of a urinal must be waterproof and nonabsorbent.

As with most types of plumbing fixtures, there are many different types of urinals. Some are mounted on floors, but most are mounted on walls. Tailpieces and exposed tubular traps are used on some models, and some have integral traps. The models with integral traps usually have a 2-inch drain outlet (Fig. 20.34), while the type with tailpiece connections have 1½ inch outlets (Fig. 20.35).

The model we will work with here is the type with a tailpiece connection. During rough-in, wood backing needs to be installed for future

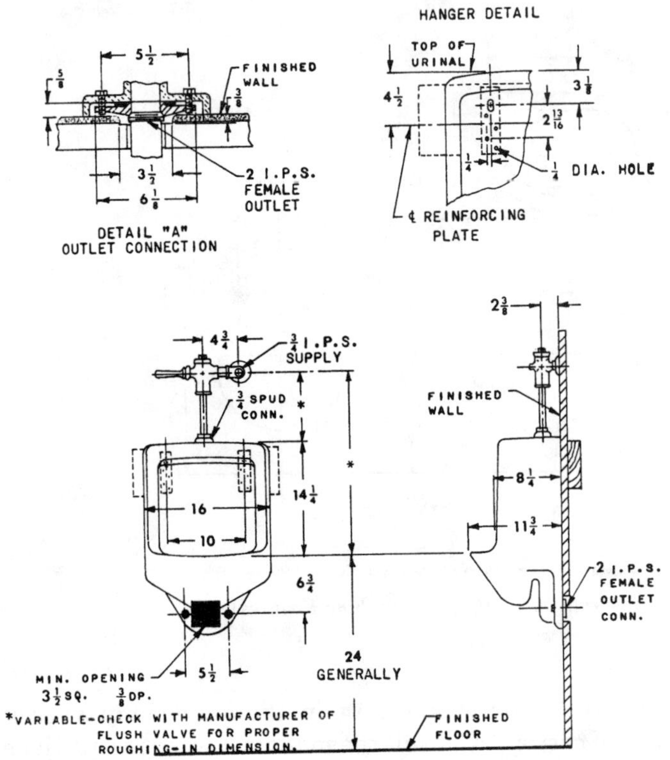

Figure 20.34 Urinal with a 2-inch drain. (*Crane Plumbing Fixtures*)

support of the fixture. One piece of backing should be installed about 19 inches above the floor and another piece should be installed about 33 inches above the floor. A 1½-inch drain should be roughed-in about 15 inches above the floor. A ¾-inch cold water supply should be roughed-in about 4¾ inches to the right of the fixture's center and about 7¾-inches above the top of the fixture. Rough-in measurements will, of course, depend on the brand and type of urinal being used. Consult a rough-in book on the specific unit being installed.

The first step in hanging a urinal is the installation of the wall bracket. In our example, the wall bracket is screwed into the recessed backing at a height of 34 inches. The fixture is hung on the bracket and two lag bolts are installed through holes near the bottom of the unit.

The flush valve is installed, per the manufacturer's recommendations, and then an 1½-inch tailpiece is attached to the drain. A trap is installed between the tailpiece and the drain, and the installation is complete.

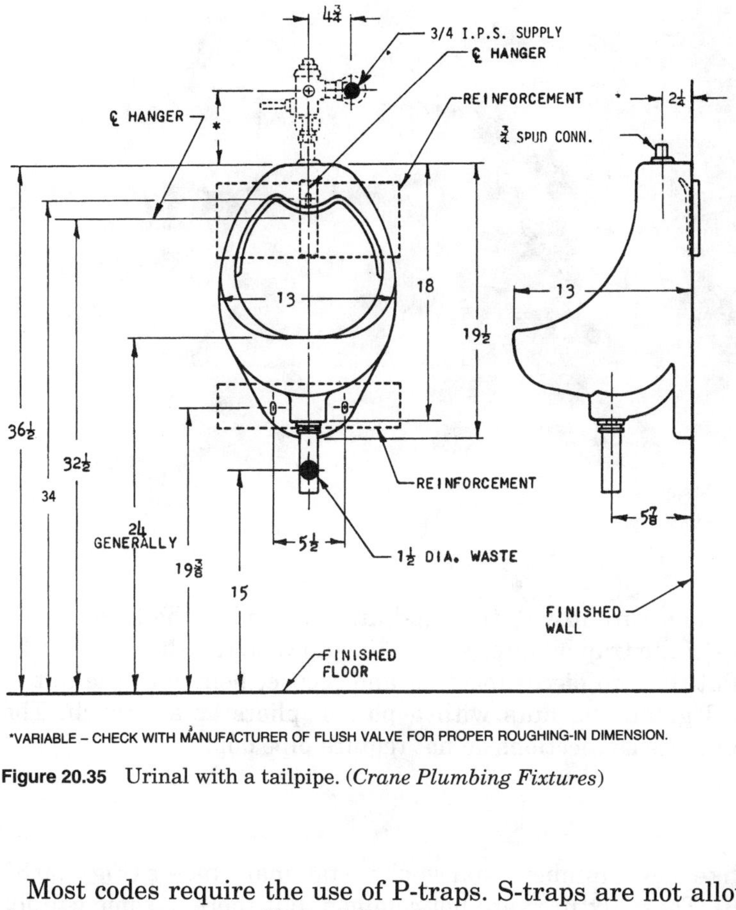

*VARIABLE – CHECK WITH MANUFACTURER OF FLUSH VALVE FOR PROPER ROUGHING-IN DIMENSION.

Figure 20.35 Urinal with a tailpipe. (*Crane Plumbing Fixtures*)

Most codes require the use of P-traps. S-traps are not allowed, and drum traps are normally prohibited, except in some combination waste and vent systems. Since P-traps are the standard in the industry, all of our installations will be done with P-traps (Fig. 20.36).

P-traps can be metal or plastic, and they come in different styles. For the sake of consistency, we will use tubular metal traps. Our trap arms will be PVC pipe.

To install a P-trap under our sample conditions, slide an escutcheon over the trap arm and glue a trap adapter on the trap arm. Loosen the nut that holds the two pieces of the trap together. Slide a slip-nut on the long tube and follow it with a slip-nut washer. Push the tube into the trap adapter. You may have to cut it to a shorter length with roller cutters.

Snug the slip-nut up on the trap-adapter threads. The fixture's tail-piece will be hanging down. Slide a slip-nut and washer up the tail-piece. Put the longer of the two ends of the U-bend of the P-trap on the

Figure 20.36 P-trap.

tailpiece, and secure it with the slip-nut. The bend in the long section coming out of the trap adapter should line up with the short end of the U-bend. Put the two pieces together and secure them with the nut at the joint. Tighten the nuts with a pair of pliers or a wrench. The threads on these connections do not require pipe dope.

Bidets

If you're like most plumbers, you won't install many bidets (Figs. 20.37 and 20.38). These fixtures are not common, but there is some call for them. The water supplies to bidets must be protected against backflow. Some bidet faucets are equipped with integral backflow preventers, but if the ones you are installing are not, you must add independent protection, usually a vacuum breaker.

The faucets for bidets can mount on a wall or on the bidet. There are several different brands and varieties available. The 1½-inch drains for bidets go down through the floor and are trapped below floor level.

Rough-in measurements vary, so check the manufacturer's recommendations for the model you are working with.

Bidets are screwed to the floor, and the drain has a pop-up assembly, similar to those used with lavatories. Read and follow the directions that come with the pop-up assembly. If you want a general idea of how to install a pop-up, review the example given in this chapter for lavatories. The drainage connections between the pop-up and the trap arm are similar to those described for urinals.

Figure 20.37 Bidet. (*Crane Plumbing Fixtures*)

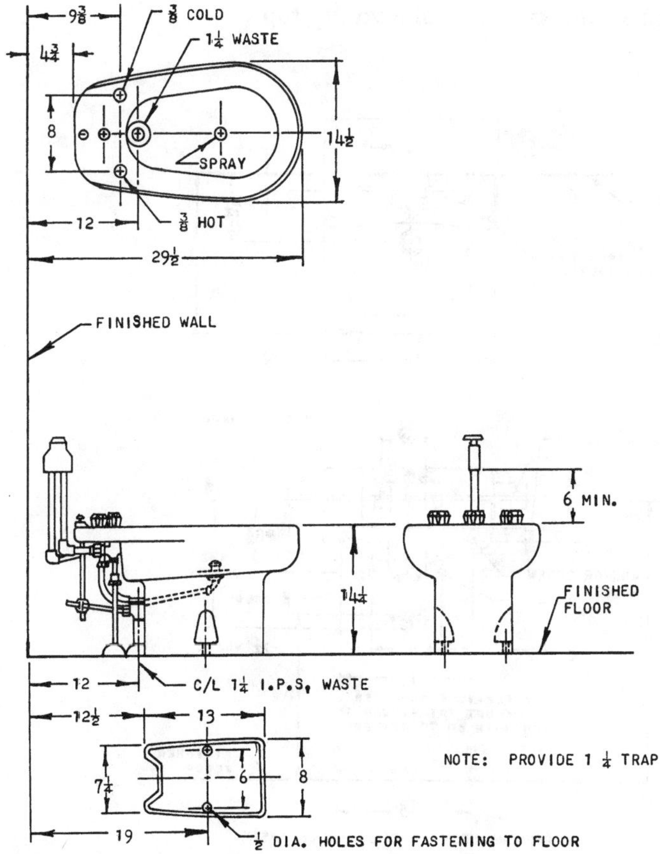

Figure 20.38 Bidet. (*Crane Plumbing Fixtures*)

Faucets for bidets can be considerably different in their installation needs. For this reason, you should read and follow the instructions supplied with the faucets you will be installing.

Lavatories

How lavatories are installed depends on the type of lavatories you are working with. There are several types: wall-hung (Fig. 20.39), self-rimming (Fig. 20.40), rimmed, pedestal (Fig. 20.41), and molded. Wall-hung lavatories hang on wall brackets that are secured to backing that is installed during rough-in. Rimmed lavatories are secured in a countertop with a metal ring and small clips. Pedestal lavatories mount on a wall, like a wall-hung lavatory, and are partially supported with a pedestal, which also acts to conceal the plumbing connections. Drop-in, or self-rimming, models are merely set into a hole in a countertop, and molded units are an integral part of a vanity top.

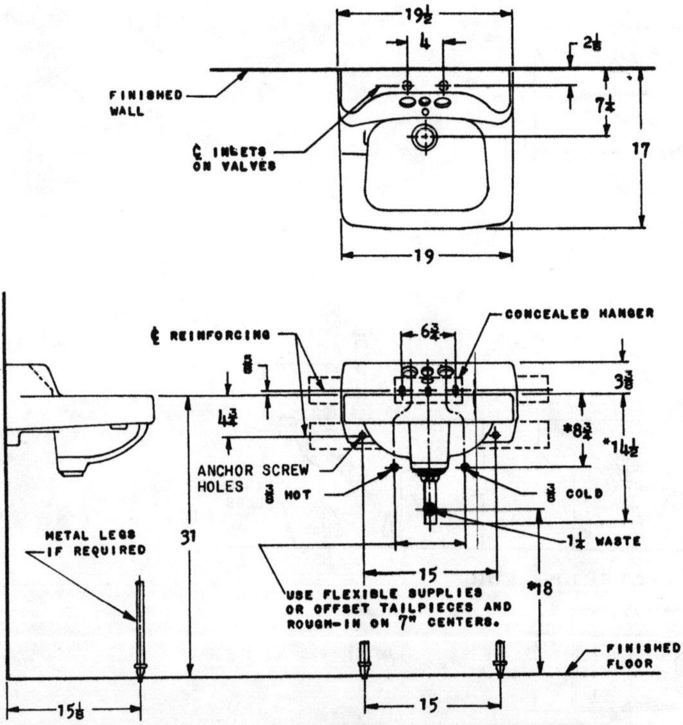

*SUGGESTED DIMENSION–VARIABLE DEPENDING UPON TYPE OF SUPPLY AND WASTE.

NOTE: RECOMMENDED USE OF ANCHOR SCREW HOLES WITH WALL HANGER.

Figure 20.39 Wall-hung lavatory. (*Crane Plumbing Fixtures*)

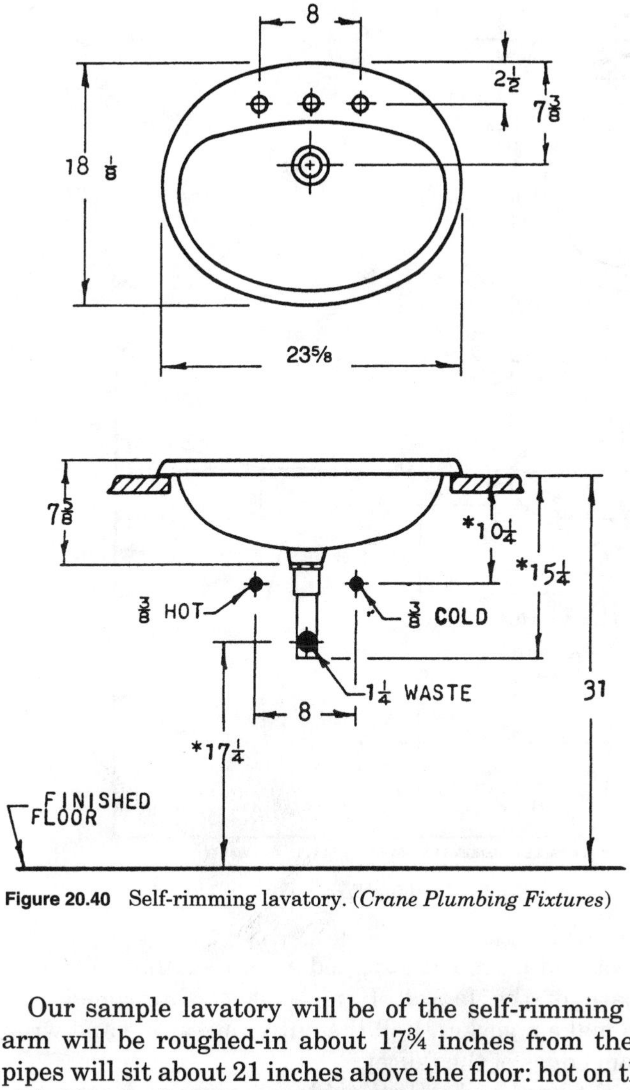

Figure 20.40 Self-rimming lavatory. (*Crane Plumbing Fixtures*)

Our sample lavatory will be of the self-rimming variety. The trap arm will be roughed-in about 17¾ inches from the floor. The water pipes will sit about 21 inches above the floor: hot on the left, cold on the right. Typically, the water pipes will be spaced about 4 inches apart. We will assume the general contractor has provided the proper hole in the counter. If the hole had not been cut, you would use a template supplied with the lavatory to mark the appropriate hole on the counter. A hole would be drilled in the counter to allow the blade of a jig saw to cut out the hole for the lavatory.

Most plumbers, at least the smart ones, install the faucets and drain assemblies on lavatories before the bowls are set into place. The work is much easier and faster this way. We are using a single-body faucet for our sample lavatory installation (Figs. 20.42 and 20.43).

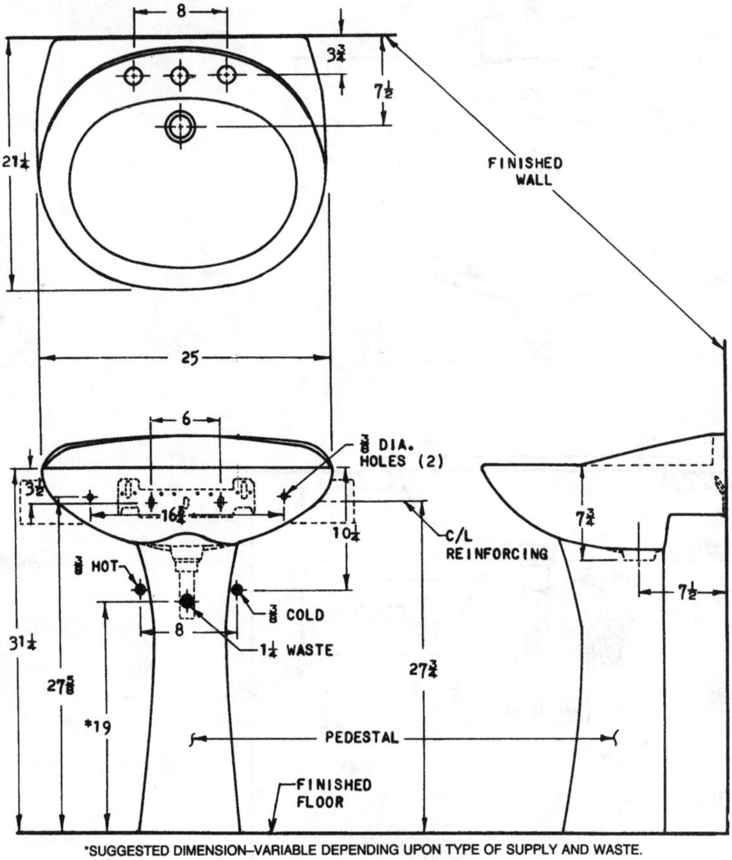

Figure 20.41 Pedestal lavatory. (*Crane Plumbing Fixtures*)

Unpack the faucet and place the supplied gasket—either plastic or rubber—on the base of the faucet. If no gasket is provided, use plumber's putty to make a gasket. Roll the putty into a line and wrap it around the bottom edges of the faucet.

Set the faucet on the bowl and slide the ridged washers over the supply stems. Follow the washers with the thin mounting nuts and secure the faucet. Install lavatory supply tubes on the bottom of the supply stems. To do this, place the beveled heads of the supply tubes in the bottoms of the stems and secure them with the supply nuts.

Moving on to the pop-up assembly (Fig. 20.44), remove the pop-up plug, and unscrew the finished trim ring. Screw the retainer nut down on the threads until it is near the pop-up assembly rod. Slide the thick rubber washer down the threads, toward the retainer nut. Install a ring of plumber's putty around the trim ring.

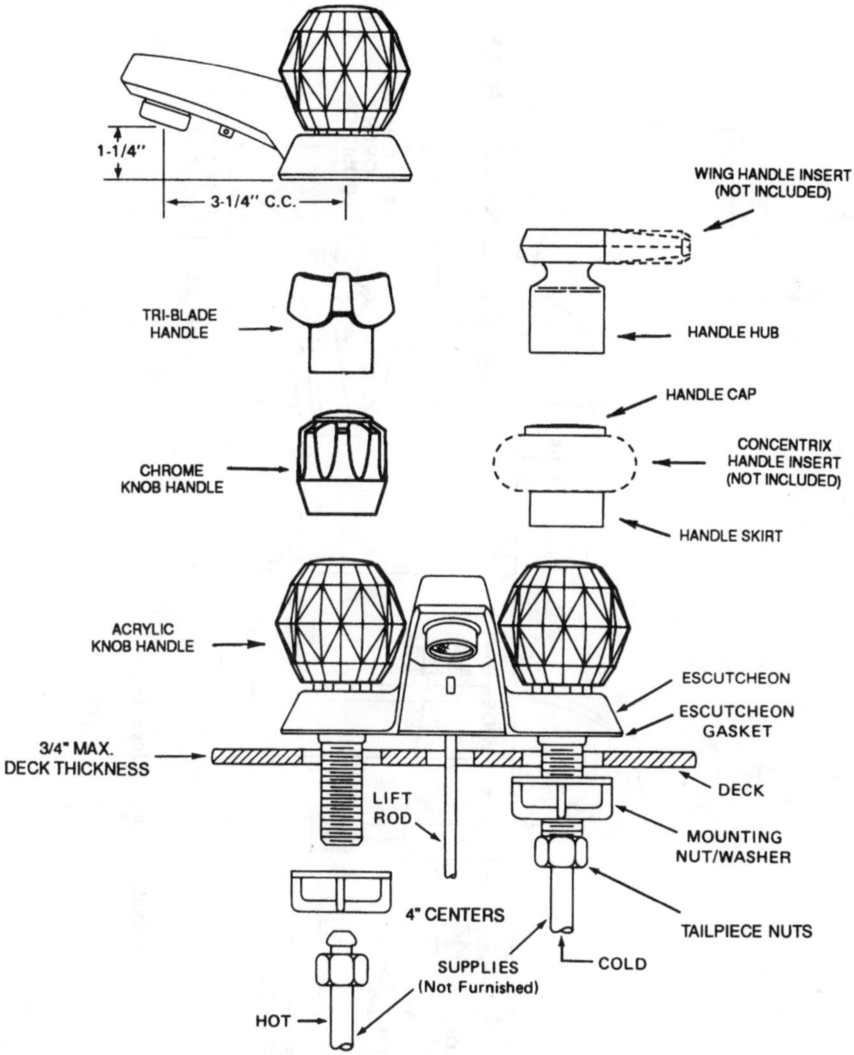

Figure 20.42 Two-handle, single-body lavatory faucet. (*Moen, Inc.*)

Apply pipe dope to the threads of the drain assembly and insert it through the drain hole, from the bottom of the lavatory. Screw the trim ring onto the drain assembly. Pull the unit down until the trim ring is in its proper position inside the lavatory. Slide the big rubber washer up to the bottom of the lavatory, bevel-side up. Screw the retainer nut up until the metal washer and nut compress the rubber washer and make putty spread out around the trim ring.

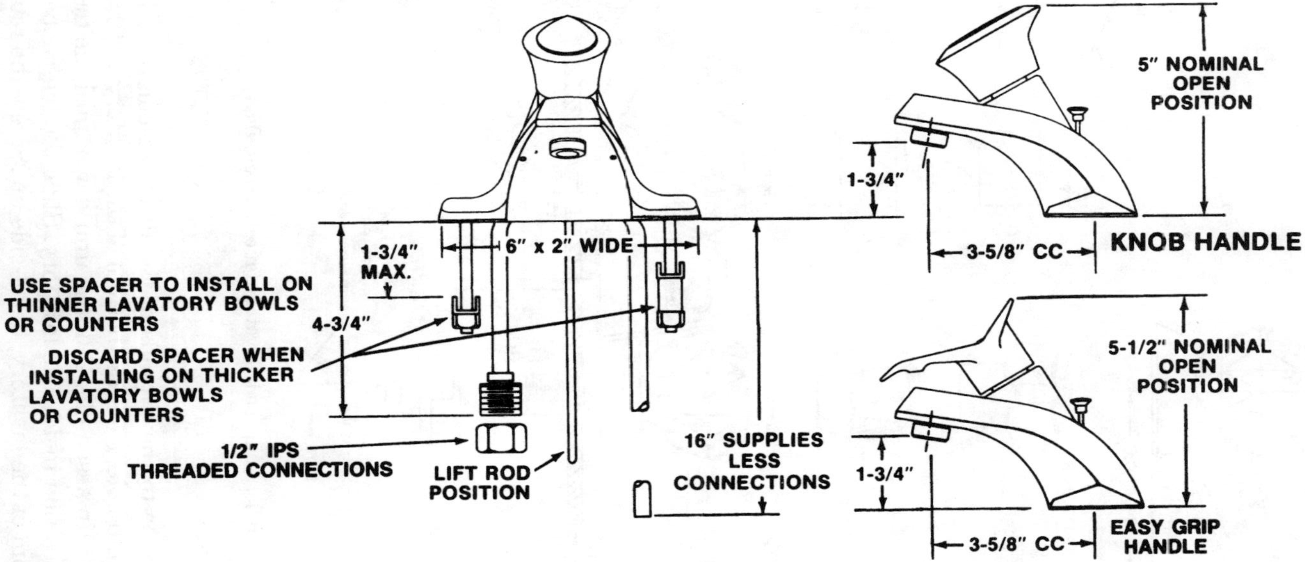

Figure 20.43 Single-handle, single-body lavatory faucet. (*Moen, Inc.*)

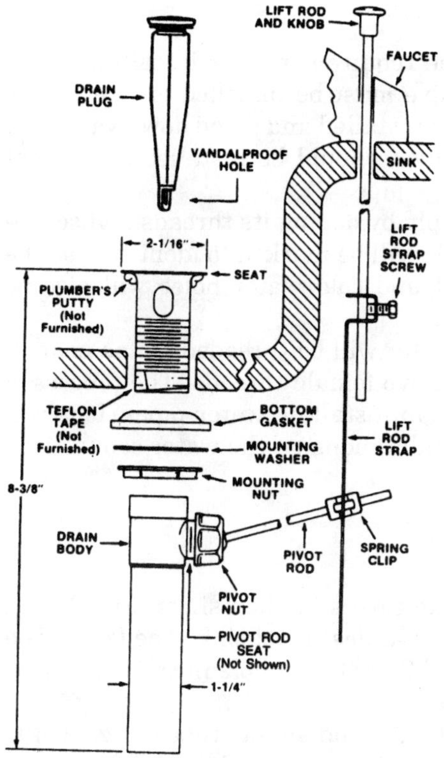

Figure 20.44 Lavatory pop-up assembly. (*Moen, Inc.*)

Now push the pop-up rod down through the center of the faucet body. This vertical rod connects with the horizontal rod coming out of the drain assembly. The connection is made with a clip and a set screw. The vertical rod enters a hole at the top of the connector strip. It is secured with a set screw.

The bottom of the connector strip will have several holes punched in its flat surface. Remove one edge of the retainer clip from the horizontal rod, and slide the rod through one of the holes in the connector strip. Replace the retainer clip on the horizontal rod.

The edge of the hole in the countertop should have a bead of caulking applied to it before the lavatory bowl is set into place. Set the bowl in the hole, and connect the waste and water lines. Dope up the threads on the tailpiece and screw it into the bottom of the drain assembly. The rest of the connections are done in the standard way with a P-trap and stop valves.

Water Heaters

We have already talked about the requirements for installing water heaters. You know a full-open valve must be installed on the supply pipe and that a relief valve must be installed and piped downward. You should also remember that no valves should be installed on the hot water pipe or on the relief discharge pipe.

The relief valve is installed simply by doping its threads and screwing it into the proper hole. The hole will be marked to identify it as the hole for the relief valve. The hot and cold water holes will also be marked.

The connections to the water heater will be threaded. Some heaters provide male threads and others have female threads. The process of installation is no more difficult than installing water pipes. It is recommended, if not required, to install unions on the water connections to water heaters (Fig. 20.45).

Kitchen Sinks

The next project we will undertake covers kitchen sinks (Fig. 20.46). Our sample sink is a double-bowl, stainless steel sink. The faucet is a single-body faucet (Figs. 20.47 and 20.48). The hole in the countertop has been cut for us.

The first step is installing the faucet and supply tubes (Fig. 20.49). This is done in the same manner as we installed the lavatory faucet, except there is no lift rod for a pop-up assembly.

Our next step is the installation of the drains. These are called basket strainers. There are different types to choose from. The trim ring will have putty put under it and will be placed in the drain hole (Fig. 20.50). A fiber washer will be installed from under the sink, and then the retainer nut or devices will be tightened. Repeat this process for the other sink bowl.

Figure 20.45 Union fitting.

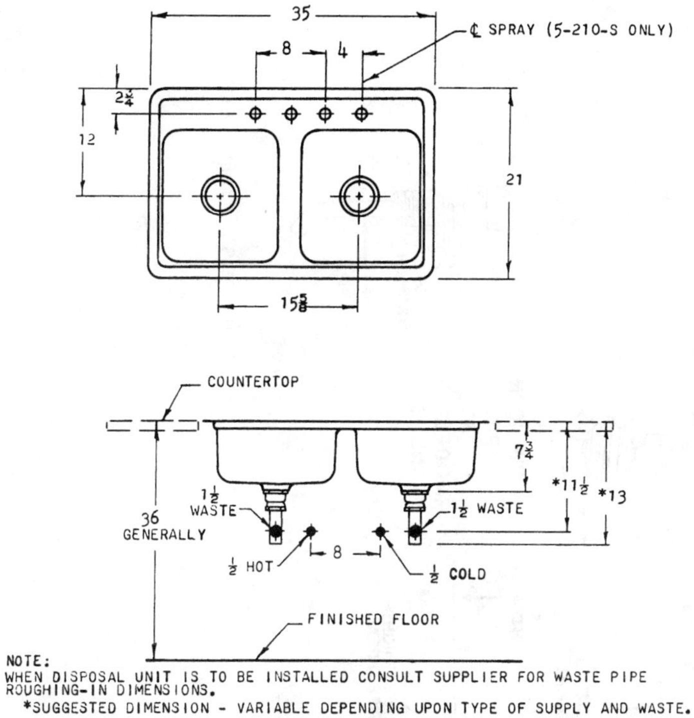

Figure 20.46 Double-bowl kitchen sink. (*Crane Plumbing Fixtures*)

Run a bead of caulking around the hole in the counter (Fig. 20.51) and set the sink in place. Slide the sink clips (Fig. 20.52) into the channel under the sink, near the edges. When the threaded portions of the clips are turned with a screwdriver, the clips will bite into the underside of the counter and draw the sink down tight (Fig. 20.53).

The trap arm will be roughed-in between 14 and 17 inches above the floor. The water supplies will be roughed-in about 20 inches above the floor, and they will be set about 8 inches apart.

Remove the slip-nuts from the basket strainers. You will find nylon washers with the nuts. Take flanged tailpieces and set the washers on the flanged surface. Slide the nuts over the tailpieces and screw them onto the basket strainers.

It is time to install the continuous waste. The continuous waste can be made of plastic or metal, and it is held together with slip-nuts and washers. These waste connectors join the two basket strainers to a common trap. They are available as end-outlet wastes (Fig. 20.54) or as center-outlet wastes (Fig. 20.55). Your choice in which to use will depend on the placement of the trap.

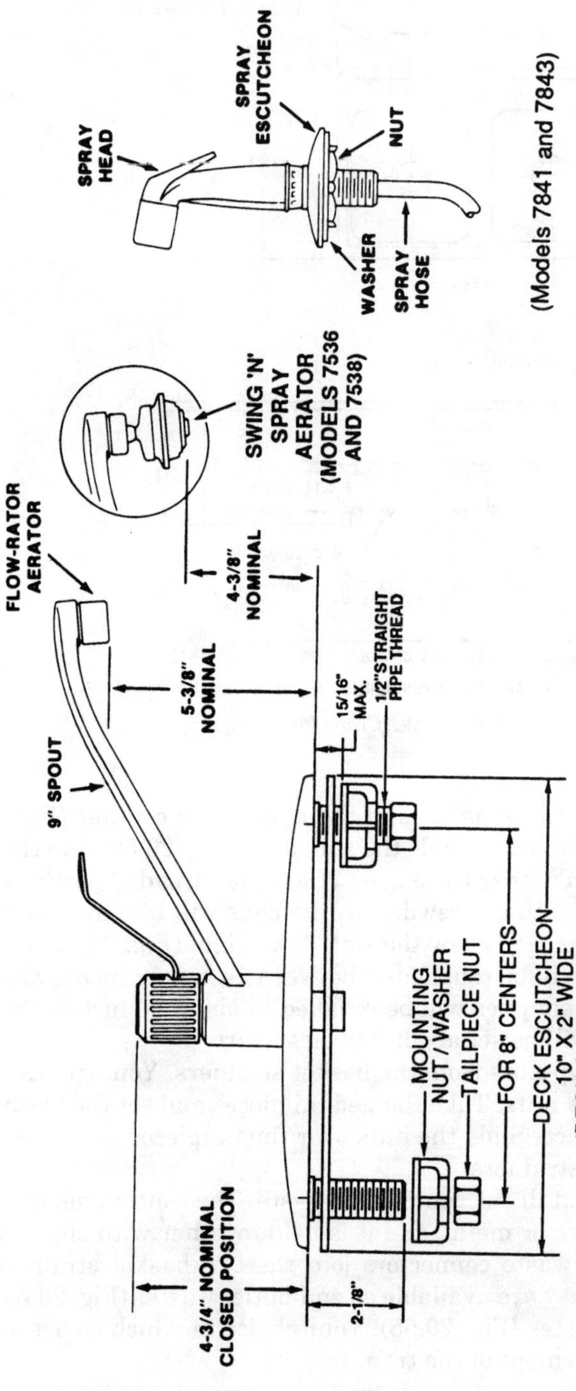

Figure 20.47 Single-handle, single-body kitchen faucet. (*Moen, Inc.*)

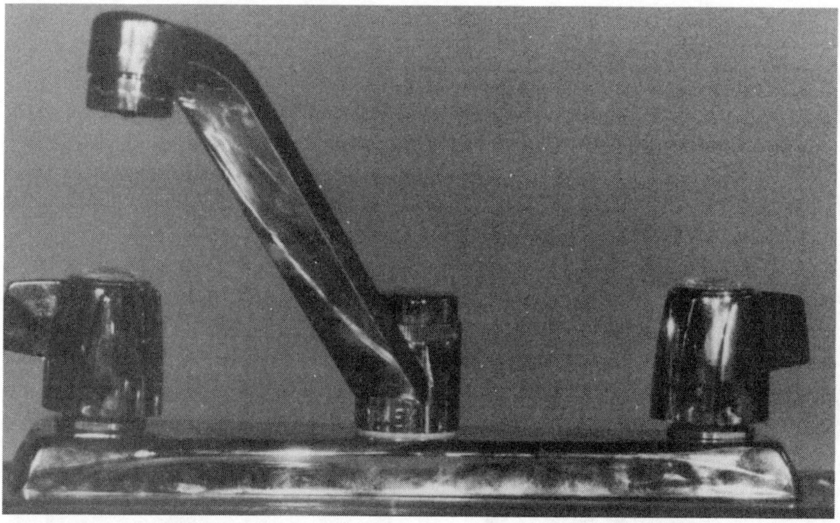

Figure 20.48 Two-handle, single-body faucet.

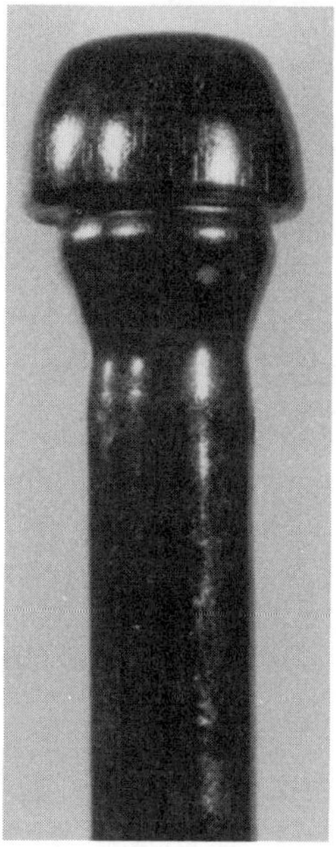

Figure 20.49 Sink supply tube.

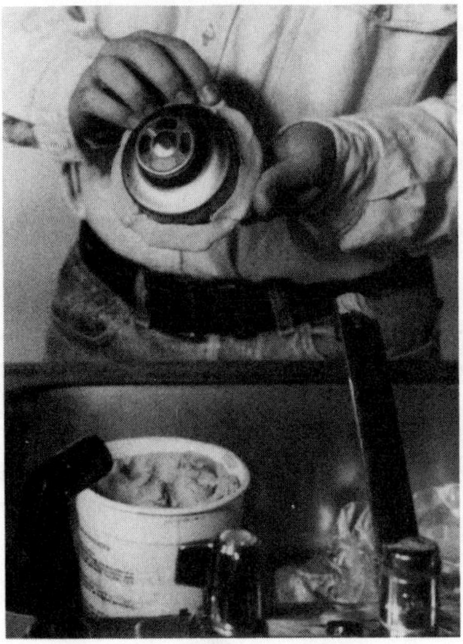

Figure 20.50 Applying putty to kitchen basket strainer.

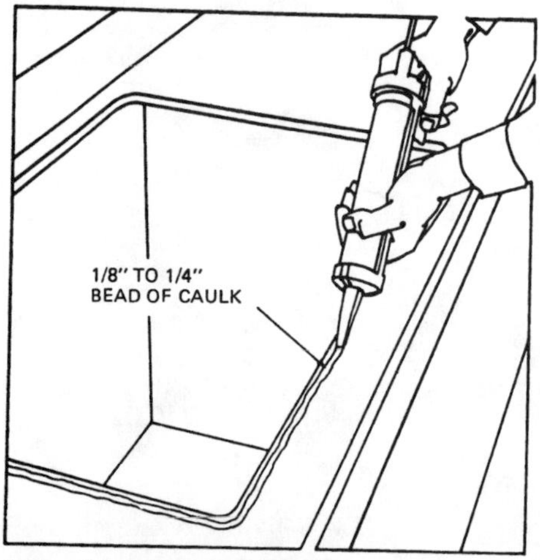

Apply a light type of caulking compound completely around all edges of sink opening. Stay approximately 1/4" from cutout. Use only a 1/8" to 1/4" bead of compound. More will make sink difficult to seat.

Figure 20.51 Caulking a sink opening. (*Moen, Inc.*)

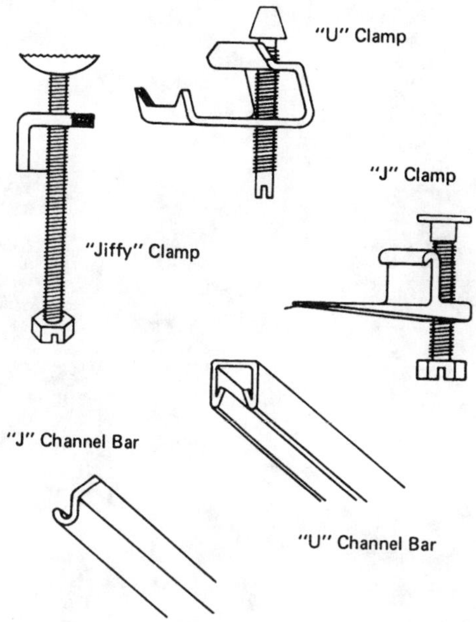

Figure 20.52 Sink clips and channels. (*Moen, Inc.*)

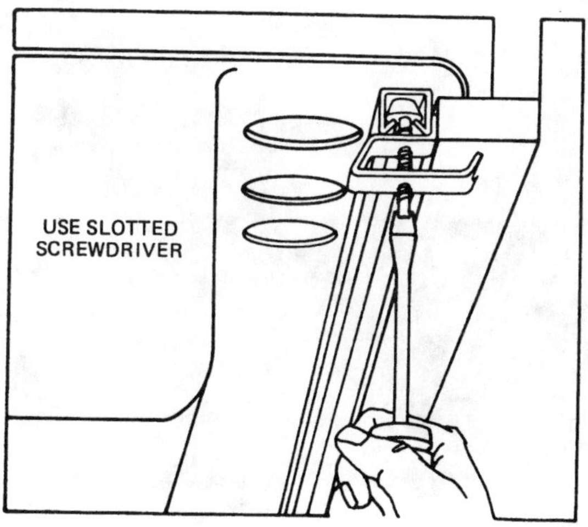

Figure 20.53 Tightening sink clips. (*Moen, Inc.*)

Figure 20.54 End-outlet continuous waste.

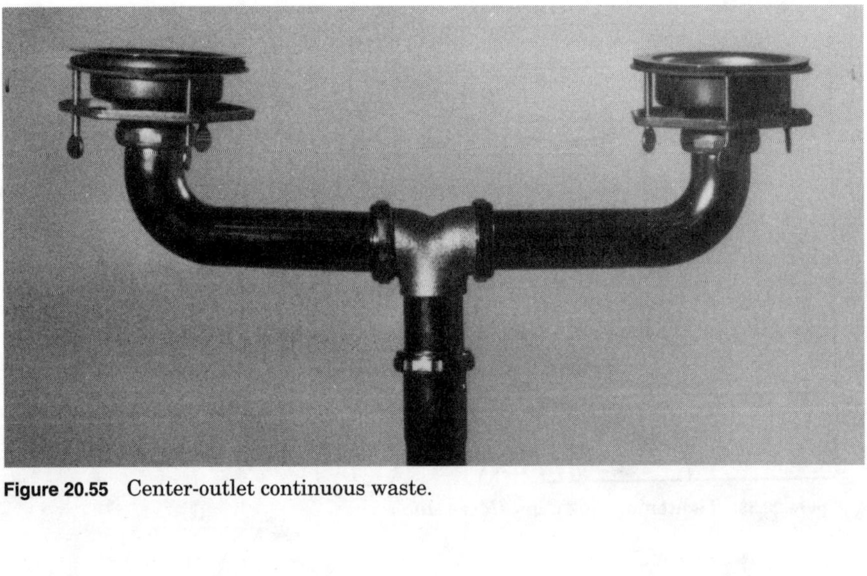

Figure 20.55 Center-outlet continuous waste.

Install the continuous waste to the tailpieces, and connect it to the trap. Install your supply stops and make your water connections. That's all there is to installing a kitchen sink.

Bar Sinks

Bar sinks are basically miniature kitchen sinks. The same installation practices are employed to install both.

Garbage Disposers

Garbage disposers take the place of basket strainers in kitchen sinks. Put putty under the trim ring and set it in the drain hole. From under the sink, slide the retaining device over the collar extending from the trim ring. Slide the snap ring into its groove to hold the retaining device in place. Tighten the three retainer bolts to make the trim ring tight.

Hold the disposer in place under the drain assembly and turn the locking flange. It will click into place and hold the disposer to the drain.

There is a small plastic elbow that installs on the side of the disposer. Remove the two screws from the retainer ring for the elbow. Put the round rubber washer on the elbow, and follow it with the retainer ring. Put the elbow in place and replace the retainer screws. Once the elbow is installed, the trap can connect to it directly.

Dishwashers

Dishwashers can be aggravating because of the limited space you have to work with, but their installation is not difficult.

The front access panel of the dishwasher can be removed by taking out the screws holding it in place. A special fitting, called a dishwasher ell, will be doped up and screwed into the water inlet opening under the appliance. Soft copper tubing will be run from the compression fitting of the ell to a three-port valve (Fig. 20.56) under the kitchen sink. Dishwashers are, of course, connected to a hot water pipe.

The drain from the dishwasher will be a rubber hose that is held on a ridged nipple with a clamp. The hose should run into the base cabinet that contains the sink and should be connected to the wye of an air gap fitting. It will slide over a ridged nipple and be held in place with a clamp. We have discussed air gaps previously.

A larger piece of rubber hose will attach to the other side of the air gap's wye branch and will deliver the waste water to the drainage system. It may be connected to the dishwasher port in a garbage disposer

Figure 20.56 Three-port dishwasher stop valve.

or to a wye tailpiece. If you connect to a disposer, knock out the thin metal plug that blocks the hole in the disposer. If you need an adapter to make this connection, they are available (Fig. 20.57).

A wye tailpiece installs on the bottom of a basket strainer or on a tubular tailpiece in the sink drainage system. It has a protrusion that accepts dishwasher hose (Fig. 20.58).

Ice Makers

Ice makers require little more of a plumber than a simple water connection. The connection is usually made with a ¼-inch tube and a saddle valve (Fig. 20.59).

Laundry Tubs

Laundry tubs can mount on wall brackets or stand on legs. The drain and faucets for these fixtures can be connected in the same general ways used for other types of sinks.

Figure 20.57 Dishwasher drainage adapter.

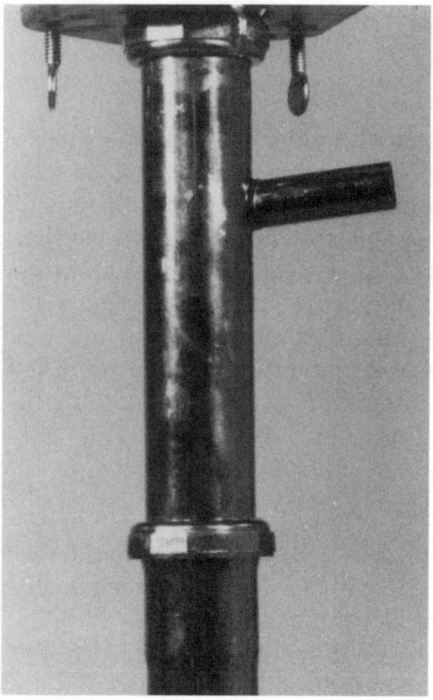

Figure 20.58 Wye tailpiece.

Figure 20.59 Saddle valve.

Handicap Fixtures

Handicap fixtures are required in some plumbing installations. Most of these occasions will occur with buildings in which the restrooms serve the public. The basic differences, from an installation standpoint, for handicap fixtures are their size and the clearances required for their installation. Of course, there are some structural differences as well.

Your local code will tell you when handicap fixtures are required in your area. It will also provide the needed information on clearances. So that you will have some idea of the differences, let me give you a few examples of how handicap fixtures differ from standard plumbing fixtures.

Handicap toilets

The seats of handicap toilets must be at least 16 inches above the floor level but not more than 20 inches above the floor. Eighteen inches seems to be the most common height of handicap toilets.

Handicap lavatories

Handicap lavatories must be installed so that the top is not more than 35 inches above the finished floor. All exposed hot water pipes under a handicap lavatory must be insulated. The faucets for lavatories may not be more than 25 inches from the front edge of the lavatory, counter, or vanity top.

Handicap bathing units

Handicap bathing units must be fitted with strong grab bars. The hoses on hand-held showers must be at least 5 feet long, and seats are often required in handicap bathing units.

There are many other code considerations for handicap fixtures, but the text in your code book on these issues is easy to understand and doesn't need to be translated.

In Closing

In closing this chapter, let me remind you that the examples given here only scratch the surface of the installation of plumbing fixtures. Because of complex designs and the countless variations in fixtures, you should always refer to the manufacturer's recommended installation instructions. Also keep in mind that code requirements vary and that it is the plumber's responsibility to be well versed in the local plumbing code. Now, let's move on to institutional plumbing and see how it differs from what we have already learned.

Chapter

21

Institutional Plumbing

Institutional plumbing is very different from residential or regular commercial plumbing. The fixtures used for institutional plumbing are sometimes unique and only used for institutional purposes. This chapter is going to make you aware of the special types of fixtures you may encounter if you venture into institutional plumbing.

Not only are the fixtures found in hospitals and other institutional facilities different, but many special requirements exist for the installation of piping. While this chapter is not going to replace your need for a local code book, it will tip you off about what to look out for. Since institutional plumbing is such a targeted area of the trade, this chapter will not take long to read.

All plumbing materials used for institutional plumbing must be approved for that purpose. Since steam traps, relief valves, backflow preventers, and other items are frequently required in this type of plumbing, much of the concealed piping that is installed must be accessible.

Institutional Fixtures

This section of the chapter will introduce you to the types of fixtures you are likely to work with when doing institutional plumbing. You will get some advice on what you can and can't do, along with an overview of the fixtures. We will start this section with a study of clinical sinks.

Clinical sinks

Clinical sinks are used for washing bedpans and flushing their contents into the sanitary drainage system. Because of their purpose, clinical sinks are often called bedpan washers. These sinks can also be used to sterilize utensils with steam or hot water.

Clinical sinks are required to have integral traps that allow the water forming the trap seal to be visible. These fixtures may work on a siphonic or blowout action, and they must be capable of resealing the trap. Flush holes in the rim of the sink are required to cleanse the side walls of the fixture. Essentially, clinical sinks must perform similarly to water closets.

As a side note, clinical sinks in soiled utility rooms may not be used as a substitute for service sinks. Conversely, service sinks must not be used as a disposal receptor for urine, fecal matter, or other human waste.

Clinical sinks must connect to the waste and vent system with the same requirements of a water closet. If the clinical sink is used as a bedpan washer, it must also be connected to a local vent. A local vent is not just a vent that is nearby. Local vents are vertical pipes connected on the fixture side of a trap that carry vapors and foul air away from the fixture. These are only used on bedpan washers.

Local vents must vent to outside air. The pipe used for these vents must have a minimum diameter of 2 inches. If the vent is venting a single bedpan washer, it may drain to the fixture being served. If a local vent is serving more than one bedpan washer, it must be drained through a trapped and vented drainage connection. Local vents for bedpan washers may not be connected to vents serving sterilizing equipment.

The trap of a local vent must maintain its trap seal. To ensure this, a ¼-inch water supply must run from the flush supply of the bedpan washer to the trap of the local vent. This supply pipe is intended to clean and maintain the trap seal of the local vent.

There are additional rules pertaining to the installation of multiple bedpan washers. Consult your local code book for a clear understanding of these requirements.

Ice machines

Ice machines may not be installed in rooms used as soiled utility rooms.

Sterilizing equipment

Sterilizing equipment is a big factor in institutional plumbing, and it requires special treatment. Any piping associated with sterilizing equipment must be accessible. Since steam is often used with sterilizers, it is important to know that the steam piping must be drained in a way to prevent any moisture from entering a sterilizing unit. Condensate from steam piping must be drained by gravity.

When pressure sterilizers are used, they must be equipped with a way of condensing and cooling the exhaust steam vapors. Sterilizers that are nonpressurized require a device that automatically controls and confines vapors within the unit.

Sterilizers, steamers, and condensers are required to discharge their drainage into the drainage system through an indirect waste, creating an air gap. If a bedpan steamer is being installed, the indirect waste pipe must be trapped.

There are a multitude of regulations on the sizing of stacks for sterilizers and steamers.

Fluid suction devices

Fluid suction devices, like the kind used by dentists, also require some special treatment. Vacuum systems must have their station receptacles recessed, visible, and readily accessible.

When a bottle-type system is installed for the collection of bodily fluids, it must be provided with overflow protection. This protection must be present at each and every outlet.

Central disposal systems for body fluids must be installed to maintain continuous service. The holding tanks must be able to be drained and cleaned while the system is in operation. In hospitals, the system must be connected to an auxiliary power station. These systems must be vented individually and directly to outside air. If the waste from these systems enters the sanitary drainage system, it must be through a trapped waste.

Piping used for central fluid suction systems must be corrosion-resistant and smooth. All piping must be accessible, and there are minimum pipe sizing requirements. These systems must be installed to produce a velocity of air flow that is not less than 5000 feet per minute.

Institutional water supplies

Institutional water supplies fall under a few code requirements that standard water services don't. For example, hospitals must be provided with two separate water services. This piping must be installed in a way to prevent a disruption in service if a water main or a water service is broken.

Common Sense Safety

Some aspects of institutional plumbing call for common sense safety measures. Most local codes require parts of the plumbing installation that are exposed to be installed in a way that eliminates the possibility of injury to patients. For example, you could not install a valve in the hallway of a hospital where a patient or gurney might bump into it. If the job you were plumbing was a mental health facility, you could not leave pipes or traps exposed, and fixtures would have to be bolted through the walls.

As I said at the beginning of this chapter, this is not all you should learn about institutional plumbing, but it is more than enough to get you started. As always, consult your local code requirements to confirm current regulations. I also promised that this chapter wouldn't take long to read, and it hasn't; we're done. Let's move into Chapter 22 and learn about special plumbing requirements.

Chapter

22

Special Plumbing Requirements

There are some special plumbing requirements that you should be aware of. Some of the topics include backflow preventers, chemical wastes, and alarm systems. But these are not the only issues this chapter will address. You will be apprised of many special conditions that require a little extra effort on your part. Like the last chapter, this chapter is short, but it is important. Now, let's move ahead with our education in special plumbing requirements.

Backflow Preventers

Backflow preventers are big issues in modern plumbing. As you have seen, there are different types of backflow protection. There are air gaps, backwater valves, check valves, reduced pressure principle backflow preventers, backflow preventers with intermediate atmospheric vents, atmospheric-type vacuum breakers, barometric loops, and pressure-type vacuum breakers. You could say there is a backflow preventer for every occasion.

How do you know when to use which backflow-prevention device? You refer to your local plumbing code. The code will instruct you on where and when to use each type of backflow preventer. For example, you will notice that a majority of the fixtures used in health care plumbing require vacuum breakers. One or two will require an air gap, and a few may be protected with either an air gap or a vacuum breaker.

If you are working in a restaurant and need to install a beverage dispenser, you will probably see that a double check valve is required and that it must have an intermediate atmospheric vent.

Should you be plumbing a building where the building's fixtures could be flooded by a backflow from the main sewer, a backwater valve should be installed in the building drain or sewer of the building. There are plenty of occasions when backflow preventers are required, and the reasons for the protection are valid.

Chemical and Special Wastes

Chemical and special wastes are an aspect of plumbing that few plumbers have to contend with. However, there may be times when the pipe you will install cannot convey its contents to a sanitary drainage system or sewer. Chemical and special wastes cannot be dumped into a public sewer until they have been neutralized.

Some of the types of wastes and debris which must be kept out of the sewer are poisons, acids, explosive liquids, grease, oil, gases, rags, ashes, and other materials that may obstruct, damage, or interfere with the intended operation of the sewer.

When you are working with chemical and special wastes, you will have to keep their drains and vents separated from the sanitary drainage system.

Alarm Systems

Alarm systems are often required on pump stations used for pumping sewage. These alarms alert people that a sewage pump is not operating properly and that the pump station may overflow. These alarms can be required for both residential and commercial jobs.

Special Pump Requirements

It is not unusual for special pump requirements to dictate the presence of two pumps in a sewer sump. These requirements are typically associated with commercial work. For example, if a day-care facility depended on a pump station to move its waste up to a sewer, it should not come as a surprise to find two pumps being required in the sump. The second pump would act as a standby. By having two pumps, the facility's waste handling requirements would not be hampered by the failure of one of the pumps.

Indirect Wastes

Indirect wastes are, in a way, a form of backflow prevention. Since there is no closed connection at these connections, back siphonage is not possible under normal conditions.

Food handling equipment is frequently required to drain through an indirect waste by way of an air gap—not an air gap device but through open air. When we talk of food handling equipment, we are not talking about average sinks; they still connect to the drainage system through normal trapped connections.

Some equipment may, at times, discharge potable water into the sanitary drainage systems. Examples of this type of equipment could include sterilizing equipment, lawn sprinkling systems, various relief valves, and so on. This type of drainage should be done through an air gap with an indirect waste.

Clear water waste that is nonpotable should also be drained through an air gap with an indirect waste arrangement. A receptor receiving clear water waste does not have to be a standpipe; it can be the hub of a fitting. However, the hub must extend at least 2 inches above the finished floor level of a waterproof floor.

Pipes draining water with temperatures in excess of 140 degrees F and steam pipes are not allowed to connect directly to a sanitary drainage system. These waste pipes must deliver their contents to the drainage system through an indirect waste.

When standpipes are required, such as for automatic clothes washers, the standpipe must be at least 18 and not more than 30 inches tall. The traps for these standpipes must be accessible for snaking the drain.

Typically, any indirect waste receptor is required to be trapped and vented. If the developed length of an indirect waste pipe exceeds 2 feet, it is normally required to have its own vent.

How big should the air gap in an indirect waste arrangement be? The distance between the flood-level rim of the receptor and the open end of the indirect waste piping should be at least twice the measurement of the indirect waste pipe. For example, if a ¾-inch pipe is used for the indirect waste piping, the air gap should measure at least 1½ inches from the rim of the receptor to the edge of the waste pipe.

When an indirect waste is not piped with an air gap, it must be installed to create an air break between trap seal of the receptor and the indirect waste pipe.

Filters and screens are often required on indirect waste receptors, and it is not uncommon for receptors to be prohibited from use in bathrooms containing toilets.

Interceptors and Separators

Interceptors and separators are used to screen and handle special wastes. These wastes might include grease, sand, oil, gasoline, and so on. Many plumbers mistakenly call grease interceptors grease traps. They are not traps; they are interceptors.

Grease interceptors are required in kitchens and food preparation areas where the waste from fixtures is likely to contain grease. However, grease interceptors are not required in residential homes. These interceptors are commonly found in restaurants, institutional kitchens, bars, and other types of kitchen areas. Garbage disposers are not allowed to be drained through grease interceptors.

Separators are installed in such places as auto repair stations. They strain and separate items like motor oil and gasoline to prevent them from entering the sanitary drainage system.

Choosing the proper grease interceptor or separator can be confusing. Normally, specifications for the devices will be provided on the blueprints for the job. However, when in doubt, consult your local code book or plumbing inspector.

Emergency Showers and Eyewash Stations

There are times when safety regulations require the installation of emergency showers and eyewash stations. These special plumbing fixtures must be provided with cold water in a way that satisfies the manufacturer's recommendations. Since these fixtures are for emergency use only, no waste connections are required.

Garbage Can Washers

Garbage can washers are not one of your everyday plumbing fixtures, but you may find occasions to install them. If you do, the water supply must be protected against backflow, and the garbage can washer must have its own trap. The waste receptor receiving the discharge from the fixture must have a removable basket or strainer that prevents large objects from entering the drainage system.

Storm Water

Storm water is handled by storm water drainage, not by the sanitary drainage system. Some areas allow storm water to mix with the sanitary drainage system but most do not. When dealing with buildings that contain more than two residential dwellings or that are used for commercial purposes, the rules for storm water drainage usually come into play. When it is necessary to provide storm water drainage, the drainage system must be designed to handle the drainage from roofs, paved areas, yards, and so forth.

Generally speaking, the materials used for storm water drainage are typically the same materials used for sanitary drainage. However, this is not always true. The rules and regulations for storm water piping vary considerably depending upon the local requirements. Because of the vast variances, I will not attempt to school you in all the requirements. I will, however, give you some examples of the types of requirements often involved.

When a roof is not pitched to drain into hanging gutters, it must be equipped with roof drains. The drains must be protected by strainers, and they should extend at least 4 inches above the roof. The inlet area of the drain should be at least twice the size of the pipe connected to the roof drain. On the occasions when the roof is used as a patio or has a similar function, the roof drains may be installed level with the roof's surface. All roof drains must be flashed to avoid leakage through the roof from around the piping.

There are roofs that have their drainage systems designed for controlled flow. These roofs may use a built-up edge, sort of a curb, to retain water and allow it to be drained at a prescribed flow. The roof structure must be designed to accept the increased weight of the water during peak loads. Typically, this type of roof will have no less than two roof drains when the roof contains 10,000 square feet or less. If the roof has more than 10,000 square feet of surface area, a minimum of four roof drains is normally required.

Pipes intended for use with storm water are not allowed to accept soil, waste, or vent connections from a sanitary drainage system. When the rain leaders, or downspouts as they are often called, are exposed to vehicular traffic, they must be protected with metal guards.

Underground storm water drains that may be affected by backflow are required to be protected with the use of backwater valves. These subsoil pipes should discharge to a trapped area drain or other approved location.

Building subdrains that are below the level of the public sewer are required to drain in a suitable sump or tank. The water collected in the sump or tank is required to be pumped out upon entry to an approved drainage system. Since the water being collected in the sump or tank is storm water, the container is not required to be vented or fitted with a gas-tight cover.

If a storm water system is allowed to comingle with a sanitary drainage system, additional rules apply. For example, rain leaders connected to a combined sewer must be trapped, as must the floor drains. Traps installed for horizontal rain conductors are required to be the same size as the pipe they serve.

Rain water drainage when connected to a combined sewer should be connected at least 10 feet downstream from any soil stack. The connection should be made with a wye fitting.

Your local code book will contain tables and text to facilitate the sizing of storm water drainage systems. The tables normally provided are not difficult to use. They deal in the total square footage of the area to be drained, the grade of the drainage pipes, the flow rate of the drainage, and the size of the drainage piping.

When sizing storm water drainage systems, you will have to know the maximum rate of rainfall that can be expected in your area. Your local code book will provide listings for the maximum rain fall in major cities.

Apprentice and journeyman plumbers are not typically required to be competent in sizing storm water systems. Master plumbers, however, are. The process can get complicated if the roof structure is not uniform. For example, if one wall of the building extends above the roof, like a facade, the vertical section of wall space must be factored into the calculations. Since the vertical wall extends upward, blowing rain that would not otherwise land on the roof could be blocked by the wall and diverted to the roof.

Because of the varying regulations on storm water drainage and the somewhat complex procedures used for sizing, we will not go into detail on the subject. As an apprentice, storm water sizing is far into your future. If you are interested in learning how to size storm water systems, consult your local code book and a master plumber. Now, let's turn to Chapter 23 and see how to test the plumbing you have learned to install.

Chapter

23

Testing Procedures

Telling you about testing procedures will not take long, but the content of this chapter is important. It is important in getting your work approved by the local plumbing inspector, and it is important to your personal safety. There are dangers involved with testing plumbing systems. You are going to be told how the simple act of removing a test ball could result in a broken wrist or worse. There are several tricks of the trade that can make your test procedures easier, and you are going to learn about them in this chapter.

Most plumbing installations must be tested and approved before they can be used legally. Not understanding what needs to be tested or how it needs to be tested can cost you time and money, not to mention embarrassment. Apprentices are usually very active in the business of testing plumbing procedures.

As an apprentice, you will probably carry water, run an air compressor, cap pipes, install test balls, remove test gear, and so on. Knowing how to do these chores properly will make you look better to your plumber, and it will make your job easier. This chapter is going to move quickly, so pay attention.

Inspections

How many inspections are you likely to have in plumbing a new house? If the house has plumbing installed beneath a concrete slab, you will have a groundwork inspection. The water service and sewer will require inspections; both of these inspections are normally done at the same time. Once the interior plumbing is roughed-in, an inspection will be required before the pipes are concealed. Then, after the fixtures are set, a final inspection will be made. Depending upon how you schedule your work and inspections, you could have as many as five inspection visits, or you might get by with just two.

Who inspects your work?

The master plumber in the company you are working with should take a look at the work. Local plumbing inspectors are normally the ones who do official inspections, but sometimes the inspections are done by private engineers.

Who schedules inspections?

The person holding the plumbing permit schedules the inspections with the appropriate local authorities: usually the plumbing inspector.

How much notice is needed for an inspection request?

The amount of notice required for an inspection request will vary from jurisdiction to jurisdiction. Most code offices require a minimum of 24 hours to arrange an inspection.

What happens if the work fails inspection?

If the work fails inspection, the defects in the work will have to be corrected, and another inspection will have to be scheduled. Most code offices require an additional fee for extra inspections.

What happens if the plumbing gets covered before it is inspected?

If the plumbing gets covered before it is inspected, the plumbing inspector has the authority to require the plumbing to be uncovered for the proper testing and inspection.

Who pays for inspection fees and permits?

The permit holder is responsible for the costs incurred to test a system and have it inspected.

Testing Building Sewers

Sewers must be tested, inspected, and approved for use before they can be covered up. They are tested from the point where they connect with the main sewer or private waste disposal system to the point where they connect with the building drain. Most code officers will allow the building sewer to be tested with air pressure or with a 10-foot head of water pressure.

How do you test a building sewer with air pressure?

To test a building sewer with air pressure, the building sewer must be capped off or blocked off with a test ball (Fig. 23.1) where it meets with the main sewer or septic tank and where it joins the building drain. Most plumbers install test-tee fittings at the junction of the building sewer and the main sewer. These fittings have a removable plug, like a cleanout, that allows an inflatable test ball to be inserted to block off the pipe.

Some inspectors will allow the final connection to be made after the test. When this is the case, the building sewer can be capped off with rubber or plastic test caps (Fig. 23.2). After the testing is complete, the inspector will watch the final connection being made and approve the building sewer. However, many inspectors want all connections made at the time of testing, and that requires the use of test-tee fittings.

How do these test balls work? Test balls are available in sizes to fit all standard plumbing pipes. In their uninflated state, the balls can be pushed into a test-tee fitting and into the opening of a pipe. A snifter valve on the end of the test ball remains accessible from the opening in the test tee. A common bicycle pump or an air compressor can be fitted to the snifter valve and used to inflate the ball.

Figure 23.1 Test ball in a test-tee fitting.

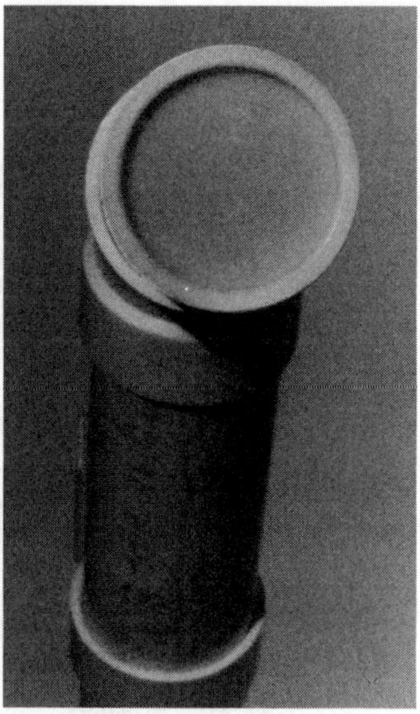

Figure 23.2 Plastic test cap.

The exterior of the test ball has ridges on it that grip the sides of the pipe when the ball is inflated. There is a large ring attached to the end of the test ball with the air valve. The ring is attached with a chain and prevents the test ball from being lost down a pipe. If the test ball were to try to fall down a pipe, the large metal ring would not go through the opening in the test-tee, preventing the loss of the testing device.

Once the section of piping you need to test is blocked off, you are ready to fill the pipe with air. Building sewers are required to have cleanouts, and the cleanout risers provide you with a place to install your test equipment.

Most building sewers have a 4-inch diameter. Snifter valves require a much smaller diameter to screw into. A series of reducers or reducing bushings will be used to reduce the cleanout riser to a size that can accept your air valve. The test rig (Fig. 23.3) will be glued onto the cleanout riser or will be otherwise appropriately attached when you are not using plastic pipe. Some plumbers, like me, use universal heavy rubber couplings to attach the test rig to the risers. This makes installation and removal a little easier.

The air valve is mounted in a test rig that contains a cut-off valve and a pressure gauge. Once the test rig is screwed into the reduced cleanout riser—pipe dope is required—you can pump the building sewer full of air. Most plumbers use air compressors for this job, but some carry tanks of air to the job site and transfer the air from the tank to the building sewer.

Figure 23.3 Test rig.

When the pressure gauge shows a reading of 5 pounds per square inch, the system is under adequate test pressure. The building sewer must be able to maintain the test pressure for at least 15 minutes. Theoretically, the plumbing inspectors are supposed to stand there and watch the gauge for the entire 15 minutes, but few do.

If the test is a success, you will be given a certificate of approval, and you can remove your test gear. The test rig will be removed from the cleanout riser and the test ball will be deflated. You will dope up the plug for the test tee and install it. When the building sewer is connected at both ends and has no openings that are not plugged, it can be covered up.

What happens if you want to test with water?

If you want to test with water, install a test ball in the same manner described for testing with air. The cleanout riser where the building sewer begins outside of the foundation will be used as the test pipe; it must rise at least 10 feet above the building sewer.

Water will be introduced into the building sewer through the test pipe. You must add enough water to have it standing at least 10 feet above the building sewer. It should be visible at the top of the test pipe. If the water level remains static for 15 minutes under the observation of a plumbing inspector, your inspection should be approved.

Once your installation is approved, you must remove the test ball and drain the building sewer. This can be dangerous. When you removed the test ball after the air test, there was little to be concerned about. However, removing the test ball while it is under a 10-foot head of water pressure can be very painful and possibly destructive to your fingers and wrist.

If you stick your hand into the clean-out tee to deflate the test ball, your wrist could be broken or you could lose a finger. As the test ball is deflated, the water pressure it has been holding back will pour through the pipe. This all happens very quickly, and you cannot afford to have your hand in the test tee when it does.

Read and heed this advice. *Never,* and I mean never, remove a test ball by putting any part of your hand into the test tee. Grasp the big ring at the end of the chain from the test tee in one hand. Use a long-handled screwdriver to let the air out of the test ball's snifter valve. When the water pressure begins to push the test ball down the pipe, pull it out with the ring you are holding onto. The first few times you attempt this, you will probably get soaking wet. However, with practice, you will be able to whip the test ball out of the pipe without getting more than a light spray of water on your body. But remember, never remove the ball with your hand in the fitting.

Test caps

Let me give you a few tips on using test caps. If you are using plastic pipe and plan to use thin plastic test caps, be prepared for leaks. These inexpensive caps don't do the best job in the world. Before installing a plastic test cap, make sure the end of the pipe you are installing it on is cut clean and square. If you try putting the cap on an uneven pipe end, it's likely to leak.

Apply glue to the end of the pipe and to the cap. Push the cap into place and then liberally coat its edges with glue. Allow the glue to dry thoroughly before testing the system. If the caps leak, you can try painting them with some more glue once the system pressure has been relieved, but you will probably be better off to cut the defective cap off and glue on a fresh one.

Rubber test caps (Fig. 23.4) that are held in place with clamps do a much better job than the plastic glue-on types do. These caps never leak when installed properly. They are much more expensive than plastic test caps, but they can be used over and over again.

What do you do if a joint is leaking?

If a joint is leaking, the first thing you will have to do is relieve the pressure from the system. Some plumbers sort of cheat by painting glue around the hub of the leaking fitting. The defective joint should be cut out and replaced. If you are testing with water, make sure the repair work is done with clean, dry materials.

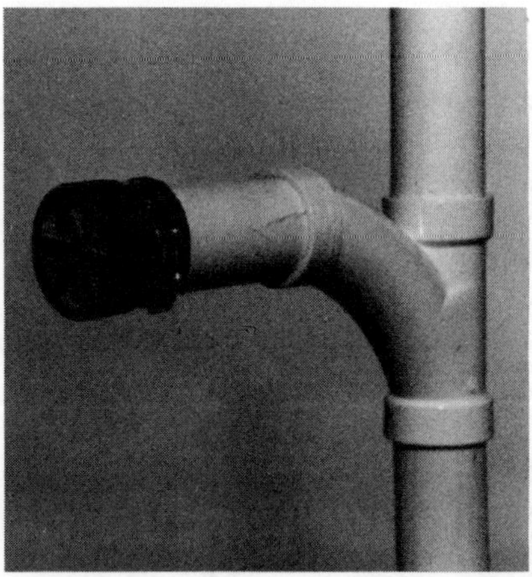

Figure 23.4 Rubber test cap.

Proper installation

Throughout this chapter on testing, proper installation is assumed. Regardless of how you apply your test, if the plumbing installation does not meet code requirements, you should fail your inspection.

Testing a Water Service

Testing a water service is a little different from testing a building sewer. If a water service is installed in one piece, with no joints between its two connection points, some inspectors will not require a pressure test; but they can.

The water service must be tested at a pressure equal to its maximum operating pressure. For example, if a water service will receive 80 pounds of pressure from the public water main, the pipe must be tested at a pressure of at least 80 pounds per square inch. Most plumbers test all of their regular water pipes with a pressure of 100 pounds per square inch.

The water service can be tested with air or water, but if water is used, it must be potable water. The water service should be capped off at the point of connection with the public water main. If air is used, it can be rigged to the opposite end of the water service with the use of reducers. The same type of test rig is used for water and drainage pipes.

If you are testing with water, you can cut the valve on at the main connection and place the pipe under street pressure. If you have to pump water into the pipe, a hydrostatic pump will be required. Hydrostatic pumps attach to a test rig that holds a pressure gauge and a boiler drain. A washing machine hose from one end of the pump is attached to the boiler drain. The valve of the boiler drain should be open.

The hose from the other end of the pump is submerged in a vessel of potable water, like a bucket. As the pump is pumped, it forces water into the water pipe and builds pressure. When the target pressure is reached and maintained, a plumbing inspector will inspect the installation and approve it, so long as the pressure gauge does not indicate any leaks. If you have a leak, let the pressure out of the system, and cut out and replace the defective joint.

Testing the Rough-In

The same basic procedures used for testing the water service and sewer are used when testing the rough-in. You can use water or air for the test. When drainage pipes are tested with air, they must hold a pressure of 5 pounds per square inch for at least 15 minutes. If a mercury gauge is used to register the test pressure, you must maintain a 10-inch column of mercury for at least 15 minutes. Water pipes must hold a pressure at least equal to their working pressure for at least 15 minutes.

If the drainage rough-in is tested with water, most plumbing inspectors require the water to fill the waste and vent system to the uppermost vent penetrating the roof. Test balls can be used to seal closet flanges (Fig. 23.5), or you can use expansion plugs for the job (Fig. 23.6). Some lenient inspectors will allow you to fill the system with water to the point of the overflow on a bathtub at the highest point of

Figure 23.5 Test ball in a closet flange.

Figure 23.6 Expansion plug in a closet flange.

the drainage system. However, both the drains and vents are supposed to be tested, and if you are allowed to terminate your test at a bathtub, the upper vents will go untested.

The Final Test

The final test of your plumbing will be done when all of the fixtures are installed and operational. In most cases, the plumbing inspectors will go from fixture to fixture, putting them into use. They will check for leaks in the water supplies and drainage fittings. Part of the inspection will involve checking to see that the fixtures are installed level and that they are secure. For example, if a toilet can be shifted from side to side, the inspector should fail your inspection.

If you are unfortunate enough to have an extremely tough inspector, you may have to put the entire system through a smoke or peppermint test. These tests expose leaks and faulty trap seals, but they are not normally required.

In the cases where they are, smoke or a peppermint oil is introduced into the plumbing system until it is seen or smelled at the vent terminals. Then the vent terminals are capped off. All fixture traps must be filled with water prior to this type of inspection. After the system is capped off, the inspector will check all traps for any trace of smoke or peppermint smell. If none is found, your work passes inspection. In about 20 years of active plumbing, I have never had to do either of these tests.

Helpful Tips

The remainder of this chapter is a compilation of helpful tips. I've already given you the basics for normal inspections; now I'm going to give you the benefit of my experience.

My first piece of advice is simple: Don't try to pull a fast one on the plumbing inspector. Experienced inspectors know all the tricks, and sooner or later, they will catch you. I'll give you a few examples of con jobs some plumbers have tried and gotten caught at.

Plumber 1

Plumber 1 decided to take a shortcut on testing his groundwork installations. This creative plumber put a plastic test cap on the end of a 10-foot piece of pipe and then set the capped end of the pipe into one of his groundwork fittings. The capped pipe was then filled with water.

To look at the system, it appeared to be under a test of water pressure with a 10-foot head. In reality, there was no water in the underground plumbing; it was all confined to the test section of pipe. The plumber did this because he was lazy. He preferred to carry only enough water to fill the 10-foot section of pipe, rather than the much larger amount needed to fill the complete system.

The con job made sense. An inspector would see the water in the test section and assume the system was filled. But one day, a smart inspector burst this plumber's bubble when he insisted that the plumber remove the end cap so that he could see the test water run out of the system. The plumber had no choice; he had to remove the end cap. When he did, no water came out, and he was caught.

You can bet your paycheck that plumber 1 had some very thorough inspections in the future. His shortcut cost him credibility with inspectors and his employer, and it made life much harder for him in future inspections when all of his work was scrutinized with a magnifying glass.

Plumber 2

Plumber 2 also tried to cheat the system. He had filled the drainage pipes of his rough-in and discovered a leak at the elbow fitting of a second-floor closet flange. Rather than drain the system and replace the defective joint, this creative plumber packed the leaking joint with wax from a wax ring. Then he painted over the wax with a heavy coating of glue. He was working with ABS pipe, so the black glue hid the wax.

When the inspector checked the job, his trained eyes noticed a slight bubble in the glue joint where the fitting had been leaking. Crawling up a stepladder, the inspector stuck a knife blade into the bubble. He found the wax, and the plumber was in big trouble.

Both of these plumbers were good plumbers who were just too lazy to do what they were supposed to do. By trying to pull the wool over the eyes of inspectors, they made their lives much more difficult. Don't try to con the inspectors. You may get away with it for a while, but when you get caught, and sooner or later you will, you will regret it.

Test balls

Don't take chances with test balls. If there is water pressure behind these test devices, they have the potential to ruin your career, your day, and your limbs. Keep your hands out of test tees.

Don't wait

Don't wait until the last minute to test your plumbing. If an inspector arrives and you are not ready for the requested inspection, hard feelings may begin to build. If you make a habit of not being prepared for an inspection, you are asking for trouble.

Don't argue

Don't argue with plumbing inspectors. Even if you can quote chapter and verse from the code book to prove the inspector wrong, don't do it. Plumbing codes are left to the interpretation of the local authorities. Since inspectors have the freedom of interpretation, you have little chance of winning an argument.

When I was young, foolish, and recently licensed as a master plumber, I challenged an inspector on the issue of an island vent. I was right, and I could prove it with my code book. When the inspector said I was wrong, I made a major issue of it. By going to his superiors I won the battle, but I lost the war.

The incident I was fighting over was awarded to me, but the rest of my inspections for the next year were very thorough. You don't want to alienate your inspectors. Even if they are wrong, give them a way out. If you put them in a bad spot, they may not forget you.

Vent caps

Don't forget to remove your vent caps after testing your rough-in. Some plumbers fail to retrieve the test caps from the roof vents, and in doing so, cripple the plumbing system. The plumbing will not work well if the roof vents are capped off.

Sewer caps

Don't stand in front of sewer caps when you remove them. If the system is under a pressure test with water, you will be drenched.

I could go on and on with old war stories, but we do have more structured learning to pursue. When you turn to Chapter 24, we will dig into the maintenance of plumbing systems. If you are ready, let's do that now.

Chapter

24

Maintenance

Maintenance of plumbing systems is required by the plumbing code. It is the responsibility of a building's owner or designated agent to maintain the property's plumbing in a safe and sanitary condition. This seems simple enough, but many plumbing systems receive no maintenance; they are neglected until repairs are needed. As a plumber, you may be asked by your customers what type of maintenance is required to keep their plumbing systems in good working order.

This chapter is not long, but it will help you to understand the need for routine maintenance and the types of maintenance that can result in fewer problems with plumbing systems. To give you a clear understanding of how to apply plumbing maintenance to a real-world situation, we will look at a property that is used for residential rental property. We will move through the property with a step-by-step check of the plumbing and outline what type of maintenance should be done.

A Typical Toilet

A typical toilet doesn't require a lot of maintenance, but there are many ways that monitoring the operation of a toilet can help avoid problems and financial losses. Let me give you some examples of the type of routine maintenance you might perform for a typical toilet.

Most plumbing maintenance doesn't require any more than periodic checks of the fixture's routine functions. What can go wrong with a toilet? A lot. It could develop a leak around the wax seal that will rot the flooring under the toilet and may ruin structural members and ceilings below the fixture. How can this be avoided with routine inspections? By flushing the toilet and watching around the base of the toilet, you might discover water seeping out from under the toilet. If you do, a wax seal that costs less than a dollar can solve the problem. Left ignored, this problem could grow into damage where the cost of repairs could be in the thousands of dollars.

Probing around the base of the toilet (Fig. 24.1) every 6 months might reveal the development of a rotted floor. The sooner you can catch this type of problem, the better off you will be.

It is not unusual for toilets to have problems with condensation. This is especially true when the toilet receives its water supply from a well. The cold water fills the toilet tank and when the warm air of the room meets the cold temperature of the water, condensation forms on the toilet tank. This can be a minor problem, but it can also be a major one.

Condensation dripping from a toilet tank can rot the flooring below it. If the problem is discovered early, it can be corrected before expensive damage is done. The tank can be fitted with insulation inserts to reduce condensation. A mixing valve can be installed to mix hot and cold water in the toilet tank, therefore eliminating the condensation problem.

Compression fittings can develop leaks if they are bumped into. Maybe when the bathroom floor was being mopped, the compression fittings on the cut-off valve or closet supply were knocked loose. Left unchecked, even a small drip can do big damage. By inspecting all connections and correcting any leaks, the life of a building's components can be extended.

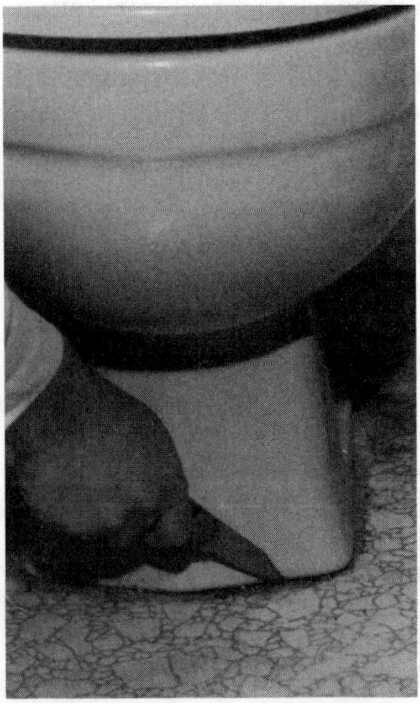

Figure 24.1 Probing around the base of a toilet.

Another frequent complaint with toilets is that they run continuously. If the ballcock or flush valve is in poor condition or improperly adjusted, the toilet may be using much more water than it normally would. Not only is this a waste of water, it can get expensive. Water pumps require electricity to pump water and cities charge property owners for every gallon of water used, whether it is wasted or not.

Tub and Shower Valves

The most frequent problem with tub and shower valves is leaks. These faucets can develop leaks that go untouched for months, sometimes years. Since the water is dripping into a large fixture, many people ignore the leak. This is a waste of water and money.

Lavatories

Lavatories can develop leaks in their water supplies, traps, and faucets. If a faucet is allowed to drip long enough, it can stain a plumbing fixture with markings that cannot be removed by normal means. When the trap begins to leak, health hazards become a factor. Leaking supply tubes and stops can rot vanities and floors. Since the water supplies and traps are typically hidden from sight, they can leak for a long time before they are noticed. However, if a routine maintenance check is made, the leaks will be found before severe damage is done.

Electric Water Heaters

Electric water heaters should be drained partially each month, but very few ever are. By opening the drain to the water heater, rust and other foreign particles are removed from the tank. When water heaters are not tended to and drained regularly, the lower heating element can become encrusted and ineffective. This runs up the cost of heating hot water and lowers the performance of the water heater.

While we are on the subject of water heaters, the relief valves on them should be tested periodically. If a relief valve becomes defective, a water heater becomes a potential bomb.

Valves

How many people ever test their plumbing valves to see that they still cut the water off? Not many. Valves can become defective over time, and they should be tested on a regular basis. You don't want to wait until water is flying all over the place to discover that your cut-offs will not shut the water supply off.

Kitchen Sinks

Kitchen sinks are a lot like lavatories. Since their plumbing is usually hidden by a base cabinet, leaks can go undetected for a long time. All fixtures should be inspected routinely.

Cracked Caulking and Grouting

Cracked caulking and grouting accounts for a lot of water damage. If water seeps into a void that is supposed to be sealed by caulking or grouting, water damage will result. When caulking and grouting become old, they should be replaced.

Trap Seals

Trap seals can dry up from evaporation when they are not used. If a fixture is not used often, it should be operated occasionally just to refill the trap seal.

Septic Tanks

Septic tanks need to be inspected every few years. The contents of the tank should be measured to determine when they should be pumped out. Failure to maintain the proper levels in a septic tank can result is backed-up drains and other problems.

Bathtub Drains

Bathtub drains frequently collect substantial amounts of hair. If routine cleaning is not performed on the tub waste, the hair build-up can restrict the flow of drainage. Enough neglect can result in a plugged drain.

The Plumber's Role in Routine Maintenance

In most cases, plumbers are not responsible for routine plumbing maintenance; the owner of the plumbing system is. However, plumbers can increase their incomes if they offer maintenance inspection services.

If you are called to a job for a routine maintenance inspection, you should do a thorough inspection. Don't just walk around and look at the pipe and fixtures. Put the system through its paces. Treat the inspection as if you were a plumbing inspector performing a final inspection on a new system. Only be more perceptive. It is more likely problems will exist in old plumbing than in new installations.

General maintenance is not a large part of a plumber's job, but it is a part of the plumbing code. Since someone has to inspect plumbing systems and perform routine maintenance, it only makes sense that plumbers should be prepared for the task.

I promised you a short chapter, and this has been short. We are now ready to move into Chapter 25 and look at the techniques and dangers of drain cleaning.

Chapter

25

Drain Cleaning

Drain cleaning is not a part of every plumber's career, but it is a specialty for some plumbing contractors. There is big money to be made in drain cleaning, and the lure of this big, and often easy, money calls out to a lot of plumbers. Dealing with overflowing toilets and backed-up drains may not be your idea of good working conditions, but some contractors wouldn't trade it for any other type of work.

There are several reasons why this type of work is so appealing to those that like it, but money is the major motivator. If I go out and replace a ballcock, I'm going to get paid for 1 hour's labor, about $37, and the parts I use. There will be some mark-up on the parts, but it won't amount to much cash.

If I go out to clear a blockage from a building drain or sewer, I'm still going to get my hourly labor rate, but I'm also going to charge a rental fee for the use of my drain cleaning machine. This fee could run anywhere from $25 to $65 dollars. For a standard machine, I'm going to charge a rental rate of $35. So I work for 1 hour and leave the job with $72. Once my machine is paid for, the rental fee is gravy. When the call comes in late at night, my labor and rental fees are going to be much higher. It is simple economics; I make more money when I can charge a rental fee on my drain cleaning equipment.

Another reason some plumbers specialize in drain cleaning is the lack of inventory needed. A typical service van can easily be stocked with an inventory worth more than $10,000. This is a lot of money to tie up for each truck on the road. If all a plumber is doing is drain cleaning, the inventory needs will cost less than $500. If I were to put five service vans on the road, this would make quite a difference.

The contents of five full-service plumbing vans might cost $50,000 not counting the cost of tools and equipment—which would include drain cleaning equipment. Putting five trucks on the road to specialize in drain cleaning could be done with only $2500 in inventory. That gives me $47,500 to use for advertising, reserve capital, or whatever I want. Are you starting to see the attraction to drain cleaning?

There are, of course, other reasons why some plumbers prefer drain cleaning over other types of plumbing. It is usually very possible for one person to handle drain-cleaning jobs. A one-plumber shop is going to have trouble doing large commercial jobs. Working alone can be tough in any type of new-construction or remodeling plumbing. Service work and drain cleaning both offer opportunity to the individual owning a business without employees.

Drain cleaning is often a fast process when the proper equipment is used. This makes it possible to get in and out of a job quickly. If you are working on a flat-rate first-hour charge, this is advantageous.

Many homeowners attempt to fix their own minor plumbing problems, but few will go to a rental center and rent a drain cleaner to auger out their drains. This makes the availability of work good for contractors specializing in drain cleaning. Even when homeowners attempt to snake their own drains, they usually fail.

When a toilet is flooding a bathroom or sewage is backed up in a bathtub, people don't put off calling a plumber. They might live with a dripping faucet or less hot water than they would like, but when their drains stop up, they call for help. This is another advantage for the drain-cleaning plumber.

As you can see, there are a number of reasons why drain cleaning is worthy of consideration as a part of your career. There are, however, certain dangers involved with drain cleaning. There are also right and wrong ways to do the job. This chapter is going talk about these facets of the job and introduce you to the tools of the trade for drain cleaning. You will also be told how, why, and where the various methods of drain cleaning are used. Now, if we may, let's get started.

Liquid and Powdered Drain Openers

Liquid and powdered drain openers that are sold in the grocery store rarely work on serious pipe stoppages. These drain openers can cut through a limited amount of grease and hair, but when the clog is tough, these over-the-counter products won't cut it.

While most of these homeowner-orientated liquids and powders won't solve the drainage problem, they do create a problem of their own for you, the plumber. The contents of these drain cleaners can be dangerous to your skin, eyes, and respiratory system. When you respond to a call for drain cleaning, ask if any chemicals or drain cleaners have been added to the backed-up pipe or fixture. It is a good idea to wear gloves, eye protection, and even a mask when clearing drain stoppages.

As a professional plumber, you will have access to some liquid drain openers that do work. Most of these liquids will have a high concentration of acid. Don't handle these materials carelessly. Also be aware that combining an acid-based product with a lye-based product can result in an eruption of harmful chemicals.

Personally, I don't like liquid drain cleaners. There are many times when even the professional-strength varieties don't work. I prefer to clean drains the old-fashioned way: with mechanical devices.

Water-Powered Drain Openers

Water-powered drain openers (Fig. 25.1) can flush out a pipe in a flash. However, when a drain is vented, getting these water-powered devices to work is a problem.

Many plumbers call these devices sewer bags. They are either canvas or rubber devices that connect to the threads of a garden hose. The device is placed in the pipe where the stoppage is and the water to the hose is turned on. The bag swells and seals off the opening of the clogged pipe. Water pressure pulsates out of the bag and builds pressure between the clog and the sewer bag. Normally, the water pressure will break up the clog or push it to a larger pipe where it can be drained out of the system. There are, however, occasions when the pressure will blow the sewer bag out of the pipe. These occasions are rare, but they do occur.

A sewer bag will only work if the water being pumped into the drain is confined between the clog and the sewer bag. If the water can rise up a vent or come out in a fixture that connects at some point to the drainage system between the bag and the clog, it will. I learned this the hard way.

Figure 25.1 Water-powered drain cleaner.

One day as a young plumber I was using a water bag to clear a lavatory drain. I had dropped the trap and stuck the bag into the trap arm. After cutting the water on to the hose, I waited for several minutes and then tested the drain. It was still clogged. I repeated this process several times, and did not realize what was happening until my father, who had gone to the truck for something, saw a steady stream of water running off the roof of the home.

The water was hitting the clog and escaping through the lavatory vent. Boy did I feel stupid. After some careful manipulation, I was able to get the sewer bag far enough into the drainage system to bypass the vent, and the drain was cleared in minutes. These devices are great if the conditions allow them to be used properly.

Air-Powered Drain Openers

Air-powered drain openers (Fig. 25.2) work along the same principles as sewer bags. Air is forced into the drainage system to blow out the clog. These devices are also rendered useless if the air is allowed to escape up a vent. Kinetic water rams are the professional version of these air-powered drain openers, but there are many homeowner versions available, including plungers.

Figure 25.2 Air-powered water ram. (*Ridge Tool Co.*)

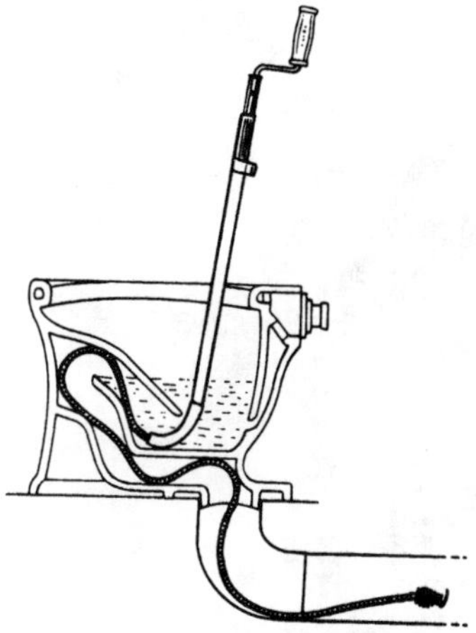

Figure 25.3 Closet auger. (*General Wire and Spring Co.*)

Closet Augers

Closet augers (Fig. 25.3) are used exclusively for clearing stoppages in the traps of water closets. Their design allows them to be snaked through the integral trap of a toilet without damaging the china, in most cases.

Most professional augers have a spring cable that is about 12 feet long. The cable is retracted into the sleeve of the auger until only the big spring head extends past the curved section of the sleeve. The curved section is set in the bottom of the toilet bowl and positioned toward the trap. As the handle of the auger is turned, downward pressure is applied on the crank. This turning and pushing motion works the auger through the trap and clears the blockage, normally.

Flat-Tape Snakes

Flat-tape snakes (Fig. 25.4) are a perennial standby. You won't find many plumbers, regardless of how much high-tech drain cleaning equipment they have, who don't keep a flat-tape snake on the truck. These snakes are available in various lengths and widths, and different heads are available for them. Before electric drain cleaners, these snakes were the cornerstone of drain cleaning.

Average lengths for these snakes are 25, 50, and 75 feet. The tape is coiled on a metal holder and is fished into a clogged drain as needed. Tape snakes work well in pipes with large diameters and few tight turns. However, they are not so effective in small drains and drains with short-turn fittings, especially galvanized drains with elbows.

Figure 25.4 Flat-type snake.

The principle behind these snakes is simple. You either break up the clog or push it through to a larger pipe. It is possible for any snake to become lodged in a tight turn or hard clog, so caution must be observed when forcing the snake into a drain.

Hand-Held Manual-Type Spring Snakes

Small hand-held manual-type spring snakes are sold in just about every store that caters to do-it-yourself homeowners. These little snakes can clear a tub or lavatory drain, but they are generally ineffective on much of anything else. The do-it-yourselfer versions are usually not of good quality and don't last long.

Professional versions of these spinner-type snakes are available for plumbers on tight budgets, but you will be better off to save your pennies for an electric drain cleaner.

Hand-Held Electric Drain Cleaners

Hand-held electric drain cleaners (Fig. 25.5) resemble hand drills with a large drum attached to them. These tools are great for sink, shower, tub, and lavatory drains. The spring cable can be fitted with different types of heads and can be made to accommodate the toughest of turns.

Electrical shocks are possible if the insulation on the wire or housing of these units is damaged. When too much cable is left hanging out of a drain, the cable can flop wildly, splattering you with black gunk from

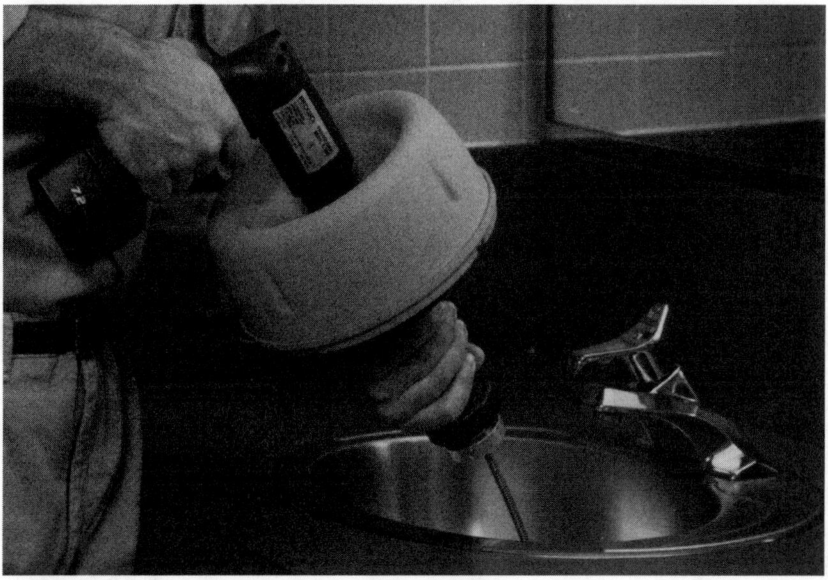

Figure 25.5 Hand-held drain cleaner. (*Ridge Tool Co.*)

the drain and causing uncomfortable bumps and bruises. Used properly, these machines are excellent for use with drains up to 3 inches in diameter.

Some models of these small drain cleaners sit on a stand (Fig. 25.6) rather than being hand held. Both types of machines work well, but most plumbers prefer the pistol-grip versions.

Big Drain Cleaners

Big drain cleaners mean big bucks. You spend big bucks to get them, and you make big bucks when you use them. These drain cleaners are used primarily for drains up to 4 inches in diameter and come in many shapes and sizes (Figs. 25.7 and 25.8). They are powerful and can hurt you badly. Most models are mounted on a frame that allows the unit to be rolled on wheels to the location of your choice.

Some models have removable drums that store the large spring cables and other models require that the cable be fed through the machine from independent rolls. I prefer the type with the removable drum. Many models are equipped with a foot control that allows the machine to be operated with the touch of a toe.

These mid-size sewer machines are available with numerous options. The price for this type of equipment starts in the $1000 range

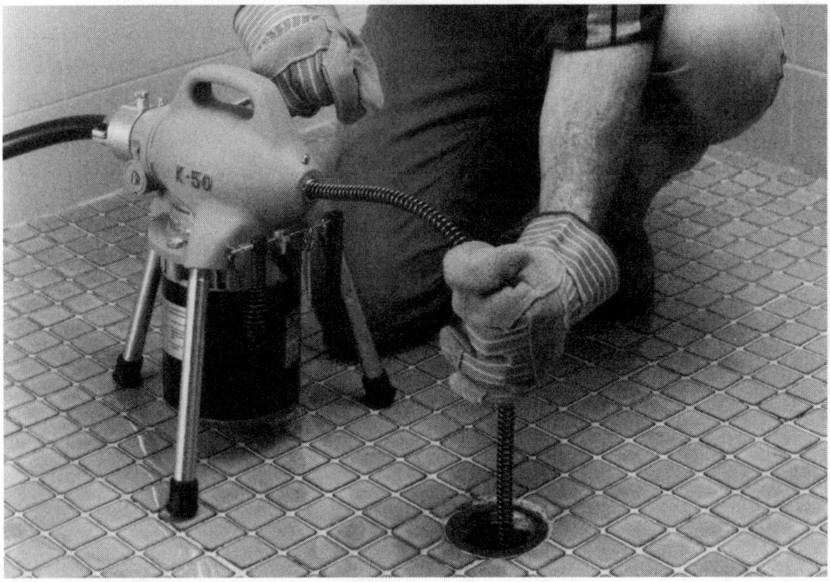

Figure 25.6 Stand-mounted drain cleaner. (*Ridge Tool Co.*)

Figure 25.7 Briefcase-style drain cleaner. (*Ridge Tool Co.*)

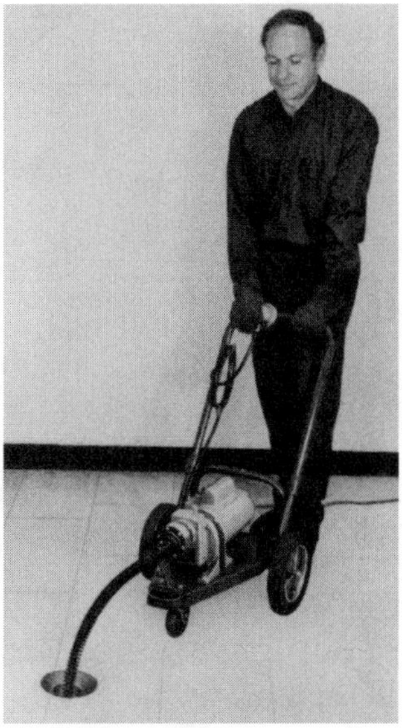

Figure 25.8 Large drain cleaner. (*Ridge Tool Co.*)

Figure 25.9 Super-sized drain cleaner. (*Ridge Tool Co.*)

and goes up. It is very important to follow the proper safety procedures when using these powerful drain cleaners.

Super-Sized Sewer Machines

Super-sized sewer machines (Fig. 25.9) are used for very large drains. If you are in commercial work, you may need this type of machine. When the drain you have to clear has a diameter of 6 inches or more, this is the piece of equipment you want.

Like other electric drain cleaners, these machines demand respect. You can easily lose a finger to the coils of a super-sized sewer machine. If you have to fight clogs in big pipes or tree roots growing in a sewer, these big bruisers will get the job done.

Well, there you have it. We have taken a look at all the most common tools and methods used to clear clogged drains. We are now ready to move on to our next and final chapter, which will teach you about troubleshooting problem plumbing. So let's turn the page and see what troubleshooting is all about.

Chapter

26

Plumbing Repairs

Plumbing repairs are a part of plumbing. Once plumbing is installed, it will need to be repaired or replaced at some time in the future. The quality of the materials used and the craftsmanship of the work will have some bearing on how soon repairs will be needed. But even the best materials and workmanship will need some attention.

We have talked at different points throughout this book about plumbers who specialize in various aspects of plumbing. There are commercial plumbers and residential plumbers. Some plumbers do drain cleaning as their major source of income. Other plumbers fill their pockets with the profits of remodeling work. Health care plumbing can provide healthy profits, and industrial work can pay off for plumbers. And then there are the service and repair plumbers.

Service and repair plumbers may never install a complete plumbing system. They might never work for more than a day on a single job. These plumbers respond quickly to calls that are unpredictable. In the morning they may be snaking the drain of a laundry tub, and by the afternoon they could be working on a fluid suction system in the local dentist office. Their work is diverse and difficult to anticipate.

A large number of licensed plumbers thrive on the variety of service and repair work. They never know from one day to the next what type of plumbing they will be doing. These plumbers never get bored with installing plumbing in townhouses where every fourth house is the same. Most of their work in done inside, in a controlled environment. They don't get a phone call first thing in the morning telling them to take the day off because it's raining.

Economic slumps and recessions have less impact on service plumbers. While people will put off adding a basement bath or building a new house, they won't live with a stopped-up toilet for long. There is much to be said for being a service plumber.

To some, service plumbers might seem to have it made. Their jobs look easy, and they are usually paid good wages. But they have their share of the dirty work. In fact, their work is often more distasteful than that of new-construction plumbers.

New-construction plumbers rarely ever deal with fixtures that are flooded with raw sewage. It is not likely a construction plumber will have to lie under a house in freezing weather to repair a pipe that has frozen and burst. The life of a service plumber isn't glamorous or glorious. It is, however, a job that requires experience.

Construction plumbers need experience, but they can get by with less on-the-job experience than a service plumber. When a construction plumber knows the plumbing code, the daily duties of the job can be met. Service plumbers must know parts of the code, but most of all, they must know how to troubleshoot and fix plumbing. They can learn much of this from reading and in the classroom, but a percentage of it must be learned with hands-on experience.

There are not many apprenticeship instructors who will prepare you for finding crumpled beer cans in the traps of toilets. They don't teach you about this possibility because they either have not experienced it or it is not a part of the prescribed curriculum. I, on the other hand, would teach you about beer cans and toilet bowls because I have experienced the problem and the frustration that went along with solving it.

The best way to learn about service plumbing is to experience it. The next best way is to talk with plumbers who are seasoned in the field of service work. The third best way is to read about it. The fact that I don't know you precludes me from being able to talk to you on a one-on-one basis, in a face-to-face conversation. I can, however, tell you with text how to troubleshoot service and repair problems, and that is what I am about to do.

The difficulty most service plumbers run into is not their inability to correct a problem, but their failure to be able to locate the problem. I have seem some mighty fine plumbers be dumbfounded by what turned out to be simple problems. These good plumbers were unable to correct simple problems because they couldn't identify the cause of the problem.

They knew what the fixtures and devices should be doing, and they knew what these same items were not doing. Where they had their trouble was in figuring out why the problem existed. I am no guru, but I've been in the plumbing business long enough that I can troubleshoot more problems over the telephone than a lot of plumbers can on the job. How am I able to do this? In one word, experience.

My experience over the years has come hard. I entered the trade on a true earn-as-you-learn basis. As a matter of fact, I was so inept in my mechanical abilities in the beginning that my father had to put the license plates on my first car for me. Now he thinks I did know how to do it, but I was really just perfecting my delegation skills. But seriously, I was not mechanically inclined as a teenager. However, I wanted to be a plumber, and I wasn't going to be denied.

As it turned out, I took to the trade very quickly and was plumbing houses on my own in less than a year. When you consider that I spent the first 6 months of my apprenticeship running a jackhammer, working a shovel, and losing 65 pounds, my proficiency in the trade came quickly.

By the time I was 21, I was a site superintendent for a 365-unit townhouse complex. I had plumbers with 30 years of experience taking orders from me, and that felt a little weird. Why am I telling you about my trek into the trade? I'm telling you my story in case you have doubts at this point about becoming a professional plumber. The trade is not easy work, but it is fulfilling. Now, let's see how you can enhance your skills with good troubleshooting and problem-solving abilities.

What Is Troubleshooting?

Troubleshooting is the act of locating problems. This may sound simple, but it is not always easy. Troubleshooting is best done through a systematic approach.

What Is Problem Solving?

Just as the name suggests, problem solving is the solving of problems. When a problem exists, troubleshooting is required to locate and identify the problem. Once the problem is found and known, problem solving is used to correct it. These two traits are inherent in solid service plumbers.

Some Examples

Let me give you some examples of real plumbing problems that I have been a part of or witnessed over the years. In doing this, you will see, first-hand, how you must learn to expect the unexpected and to look beyond basic plumbing concepts.

Problem number 1

Problem number 1 describes the time I found a beer can in a toilet. I was called to a rental property because the toilet was overflowing. After arriving, I used a closet auger in an attempt to clear the toilet's trap.

The auger would not go through the trap. There are times when augers have difficulty getting through the integral trap of a toilet, so I accepted the fact, after many attempts, that the auger just couldn't find its way through to the drain pipe.

I removed the toilet from its flange, with the intent of snaking the drain. When I noticed there was no water standing in the drain pipe, I became suspicious. This was the first step of my successful troubleshooting. Some plumbers would have snaked the drain, even though no water was visible.

I filled a 5-gallon bucket with water and poured it down the drain. The water went down the pipe without hesitation. All of a sudden I had eliminated the possibility of the problem being in the drain and could point my finger at the toilet.

With the toilet laying on its side, I inspected the drain outlet. There were no visible obstructions. When I attempted to auger the toilet from the drain outlet, the auger still balked. Being young and inexperienced, I was confused.

As time went on, I was ready to take a hammer to the toilet, just to see what was wrong. However, I didn't want to have to buy the customer a new toilet, so I continued to hunt and peck for an answer.

After going out to my truck for a wire coat hanger, I tried to fish the coat hanger through the toilet. It went through after a little artful manipulation. When the wire was in the trap, it was easy to hear that it was hitting more than the porcelain fixture. I couldn't retrieve whatever was in the trap.

I called the landlord, and he met me on the job. With his permission, I broke the toilet with my hammer. We found a beer can in the trap. The can appeared to have been put on the floor and stepped on. The compressed can had somehow found it's way into the trap, but it couldn't get out. At least not until I broke the toilet into bits.

The landlord had to buy a new toilet, and I suspect the tenant was charged for it. In any event, I got paid and had learned my first nonconforming lesson about service work.

Problem number 2

Problem number 2 has to do with a toilet that wouldn't flush properly, due to a code violation. This problem was being dealt with by one of the plumbers I supervised. The plumber had years of experience with service work, but this job perplexed him. The plumber had responded to a call for a toilet that was sluggish when it was flushed. The homeowner complained that the toilet had always flushed slowly and that it frequently came close to overflowing.

The plumber went through all the normal motions of clearing the toilet's trap and drain. But all of his efforts failed. Assuming the toilet was defective, he sold the homeowner a new toilet. It was late in the day by the time he installed the new toilet. When he tested the new fixture, it suffered from exactly the same problem the old one had. Confused and embarrassed, the plumber called the office for help.

I was in the office when the call came in. In talking with the plumber, it seemed he had not left any stone unturned. Since I couldn't solve the problem over the phone, I drove out to the job. The homeowner was understanding and kind. My plumber was beet-red from selling the lady a toilet she didn't need.

I flushed the toilet twice and then went into the basement. As I neared the closet flange and saw the piping arrangement, I was relatively sure I had discovered the problem. The experienced service plumber had seen the same piping setup but had paid no attention to it.

The pipe coming out of the closet flange dropped straight down for several feet. It went into a short-turn elbow fitting that turned to a poorly graded horizontal pipe.

I was convinced that the when the toilet was flushed, the water was rushing down the vertical pipe and bouncing off the interior of the fitting. The combination of the bounce and the poorly graded pipe would not allow the toilet to flush properly. The rapid descent of the water, in my opinion, was leaving toilet paper stranded in the pipe, further complicating the problem. My plumber argued with me, in private, that what I was suggesting couldn't cause the problem.

Since he had tried every other option, he consented to humor me. We cut out the long vertical drop, which was in violation of the plumbing code, and reinstalled the piping with a shorter drop and 45-degree fittings. Once the joints had cured, we tried the toilet several times. It worked perfectly.

I offered to reinstall the old toilet and discount the labor charges for my plumber's time in not solving the problem. The homeowner accepted my offer for the discounted labor, but agreed to keep the new toilet.

It has been over two years since we corrected that problem, and it has never reoccurred, proving I was right. Not only was I right, it was glaring proof that the plumbing code is not just a bunch of hogwash. If the pipe had been installed to code, the problem would never have existed.

My plumber spent about 6 hours trying to solve a problem that I solved in less than an hour. What made the difference? My experience, knowledge of the plumbing code, and the fact that he had already exhausted all normal means of troubleshooting. Could I have solved the problem quicker even without his prior elimination of other problems. Probably so, but it might have taken more than an hour.

Problem number 3

Problem number 3 involves a wall-hung toilet that would not flush properly. This toilet would not empty its contents. The water would swell up to the flood level rim and threaten to pour over before slowly subsiding.

Two journeyman plumbers tried to fix the toilet on two different days. Neither was successful, and the last plumber decided to replace the toilet. The toilet was replaced and the problem continued to exist.

The plumbers involved in this problem worked for a company I was consulting with. I happened to be in the office when they came in telling their boss about the problem with the toilet. They claimed to have tried everything, but nothing worked.

Upon hearing the symptoms, I asked how long the toilet had been acting up. One of the plumbers didn't know, and the other one said he thought it had never been right. Based on the assumption the toilet had never performed properly, I asked if either of the plumbers had checked the fitting behind the carrier bracket.

They told me they had snaked the drain, even the vent, and had replaced the old toilet with a new one, but neither man had checked the carrier fitting. I suggested a check to see if the carrier fitting was a sanitary tee or a short-turn elbow.

One of the men remembered the fitting being a sanitary tee with a cleanout plug in the end of it. I told them to replace the sanitary tee with a combination wye and eighth bend fitting. They did, and the problem was solved.

When the toilet was being flushed, its contents were hitting the back of the short-turn tee and being forced back toward the toilet. The long-sweep fitting solved the problem.

This job cost the company's owner a new toilet and several hours of labor, at a rate of $37 per hour, to satisfy the customer. There was no way the customer was going to pay the full bill for the plumbers' incompetence. The plumbers had overlooked a simple problem, but they aren't likely to make the same mistake again.

Experience is gained in the field from jobs like these. All three of these problem jobs taught the mechanics something new about service work. I could give you dozens of other examples, but my point has been made. Before you can fix a problem, you have to find and identify it.

When you attend classes as an apprentice, many instructors will teach you how to correct problems, but most of them won't show you how to find the problem. The instructor may not have the hands-on experience to tell you what to look for. Time could be a factor in your not being taught how to find the problem. Classes are typically scheduled with a tight budget on time, and the instructor might not have the opportunity to teach you troubleshooting techniques.

If you leave your apprenticeship class with the ability to correct the problem, you have half the battle won. However, until you determine where and how to apply your repair knowledge, the skills learned in the classroom will do you little good. Let's look now at how you can decide where and when to use your skills.

Troubleshooting Toilets

Troubleshooting toilets can be tricky, but it usually isn't. This section of the chapter is going to give you advice on what to look for when a toilet is acting up. Let's get started with a toilet that will not flush.

The toilet won't flush

You're on a job and the toilet won't flush. It's not that the water backs up in the toilet when being flushed; it just won't flush. What might be wrong that would prevent the toilet from flushing?

The first thing to do would be to remove the cover of the flush tank and see that it is filled with water. If it's not, check the cut-off valve on the toilet supply and make sure it is open. Assuming there is water in the tank, check the connection between the toilet handle and the flapper or tank ball. If the handle is not lifting the covering device from the flush valve, the toilet can't flush.

The toilet will attempt to flush

Let's pretend the toilet will attempt to flush but flushes poorly. What could cause this type of problem? The drain pipe could be blocked with a stoppage. A partial stoppage would allow the water to go down slowly, but a full blockage would force the waste to overflow from the lowest available opening. If the plumbing system dumps into a septic tank, the septic tank might be the problem; it could need to be pumped out.

What else could make a toilet sluggish in its flush? The vent for the toilet could be obstructed. A blocked vent can make any fixture drain slower.

The flush holes around the rim of the toilet bowl could be clogged. This is a problem that baffles many inexperienced plumbers. There are a series of holes around the inside rim of a toilet. The holes are not visible looking down into the toilet. They can be seen by holding a mirror in the bowl and pointing it to reflect the underside of the rim.

These holes sometimes become blocked with mineral deposits, especially when the water supply is coming from a well. If the holes are blocked, the fixture can't flush normally. Sticking the end of a coat hanger in the holes will break up the crusty mineral deposit and free the holes.

The water level in the flush tank could be too low for a good flush. Some people put bricks, jugs of water, and other assorted items in their flush tanks to reduce the amount of water used when flushing the toilet. A few handy homeowners even adjust the ballcock to lower the volume of water in the tank. If the amount of water in the flush tank is reduced too drastically, the toilet cannot flush properly.

Flappers (Fig. 26.1), tank balls (Fig. 26.2), and other devices covering the outlet of the flush valve could be causing the trouble. If the covering device does not stay up long enough, not enough water is allowed through the flush valve. This results in a poor flush. Adjusting the chain or lift wires connected to the toilet handle will usually correct this problem, but it may be necessary to replace the covering device.

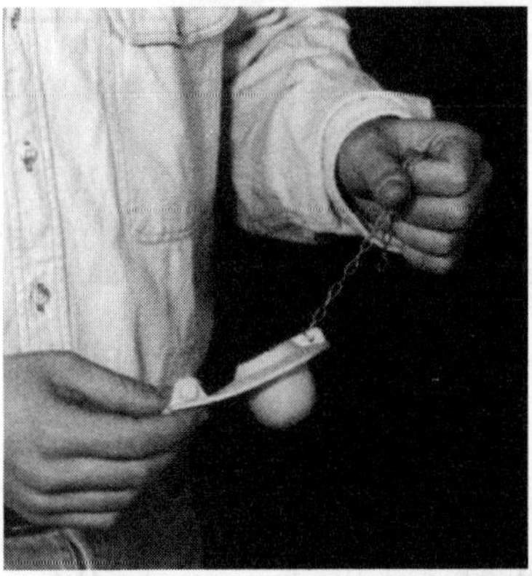

Figure 26.1 Flapper for toilet tank.

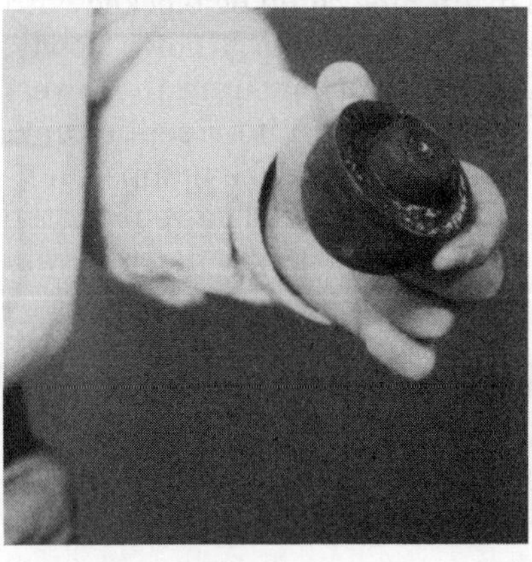

Figure 26.2 Tank ball for toilet tank.

Some toilets are finicky. If the refill tube from the ballcock is not in the refill tube of the flush valve, the toilet bowl will not receive the amount of water it was designed to use. This could cause the toilet to flush abnormally. Any of these problems could result in a toilet that flushes slowly.

Water dripping from the flush tank

If you walked into a bathroom and saw water dripping from the flush tank, what would you think? I have seen plumbers condemn a tank as being cracked, when the problem was condensation. Many toilet tanks sweat in the summer. The cold water in the tank reacts with the hot air in the bathroom to form condensation on the outside of the tank. If you see a tank with beads of water on it, think about condensation.

If you rule out condensation, check the tank-to-bowl bolts. If these bolts have rusted or become loose, they could be allowing water to drip.

The tank-to-bowl gasket could contribute to water dripping from the flush tank, and the nut holding the flush valve in the tank could be loose or cracked. And, of course, the tank could be cracked.

The flush tank is empty

Suppose the flush tank is empty. What might cause a toilet tank to be unable to fill with water? The supply valve could be closed. The valve may have been closed accidentally while cleaning, or a curious child might have turned the handle. It may seem unlikely that a homeowner would call a plumber before checking the valve, but I've had it happen.

If the temperature has been or is below the freezing mark, the water supply may be frozen. This is an especially good possibility if the pipe is installed in an exterior wall. It is also possible that a pipe is broken somewhere, depriving the supply tube of water.

Defective ballcocks can result in toilets that will not fill. You can tell if water is getting to the tank by loosening the ballcock nut and seeing if water sprays out from the opening.

A slow water supply

Any of the following problems could cause the flush tank to fill slowly: a partially frozen pipe, a partially closed supply valve, a partially obstructed supply pipe, a defective ballcock, or just plain low water pressure.

A roaring ballcock

Have you ever heard a roaring ballcock? If someone is complaining of a toilet that makes a lot of noise when it is flushed, go directly to the ballcock.

Seeping water

Seeping water around the base of a toilet usually indicates a wax ring that needs to be replaced. However, the toilet bowl could be cracked.

A running toilet

When the toilet runs constantly, you have many options to check for the cause. If the water supply fails to cut off after a flush, check the flapper or tank ball first. I've gone to houses and found the chain that connected the flapper to the handle tangled. This causes the flapper to stay open and keeps the water running. Don't overlook the obvious.

If the covering device on the flush valve is seating properly, you will have to look a little further. Start by cutting off the water supply. If the water level is above the refill tube of the flush valve, your problem is not in the flush valve or its accessories. When the water level is below the refill tube and the water is running continuously, the water must be leaking through the flush valve. There is one exception to this rule. If the refill tube on the flush valve is cracked, it can fool you.

The covering device on the flush valve may be defective. It is also possible, if the flush valve is a metal one, that the flush valve has become pitted and is allowing water to seep past its seal.

When the problem is related to the water supply, check the ballcock (Fig. 26.3). If the ballcock is equipped with a float rod and ball (Fig. 26.4), check these items; sometimes a float fills with water.

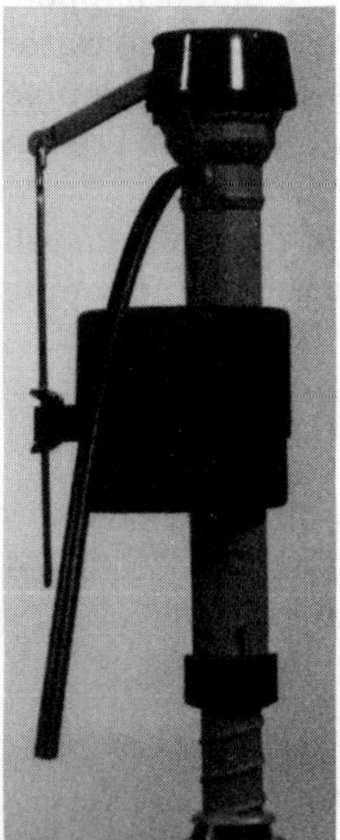

Figure 26.3 Vertical ballcock.

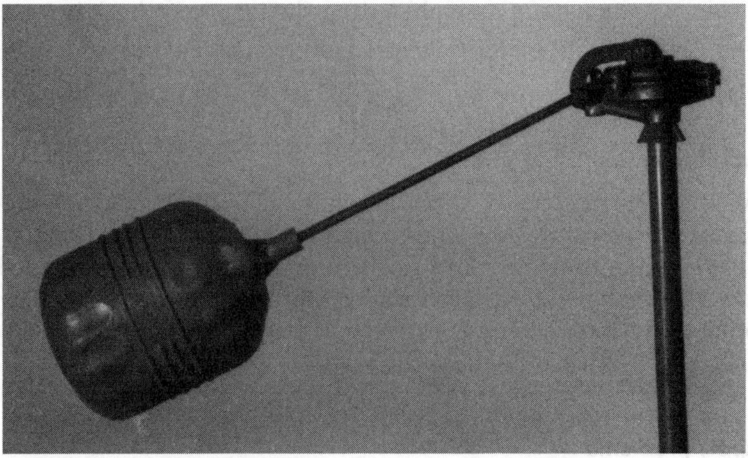

Figure 26.4 Float-ball ballcock.

Troubleshooting is nothing more than a process of elimination. You work your way through the possible problems until you find the real problem. The key is in learning what to look for first, so you don't waste time looking for the wrong types of problems. Now let's divert our attention to lavatories.

Troubleshooting Lavatory Problems

Let's talk about troubleshooting lavatory problems. These problems could be centered around the drains, water supplies, or the faucet. We will start this section on the subject of lavatory drains.

Lavatory won't drain

What will you do when a lavatory won't drain? The most logical place to start is with the trap, unless you notice that the pop-up drain is not operating. Remove the trap and check it for a blockage. If the trap is clear, you can plan on snaking the drain.

Lavatory drain is sluggish

What could the problem be when a lavatory drain is sluggish? The pop-up plug should receive your first attention. Make sure the plug is working properly. If it is, remove it and check for a build-up of hair. Pop-up plugs and lift rods often collect gobs of hair.

If you rule out the pop-up assembly, drop the trap and check it for a partial blockage. The remaining options would be a partially clogged drain or a blocked vent.

Bowl won't hold water

If the lavatory bowl won't hold water, check the pop-up plug. It may not be seating properly, or the pop-up seal may be bad.

It is also possible that the putty seal around the trim ring is inadequate and is allowing water to leak under the ring and into the drain. This problem drives some plumbers crazy. If you don't know to look for it, you may never figure out where the water is going.

Water supply problems

What are some of the water supply problems you might run into? If you don't have any water coming to the fixture, check for closed valves, frozen pipes, and broken pipes.

If you've got water but not much of it, check the aerator. Aerators frequently become clogged and slow the flow of water into the lavatory bowl. If the aerator is not the problem, check for partially closed valves, partially frozen pipes, and partially blocked pipes and supply tubes.

Other considerations might include checking the faucet for an obstruction. It is possible for mineral deposit, sand, rust, or other foreign objects to become lodged in a faucet.

General low water pressure could be the problem. Check other fixtures to rule out the possibility that the water pressure is low throughout the system.

Lavatory faucets

Lavatory faucets come in all shapes and sizes, but they share many of the same possible problems. If the faucet is dripping from its spout or will not shut off, you can assume it needs parts replaced. The parts might be washers, seats, ball assemblies, or cartridges, depending upon the style of the faucet.

If the faucet is leaking around the base of its spout, there is probably an O-ring that needs to be replaced.

Lavatory faucets share these potential problem characteristics with other types of faucets. Whether you are working on a kitchen, laundry, or tub faucet, the same basic principles apply. You will need a basin wrench (Fig. 26.5) for some of these repairs.

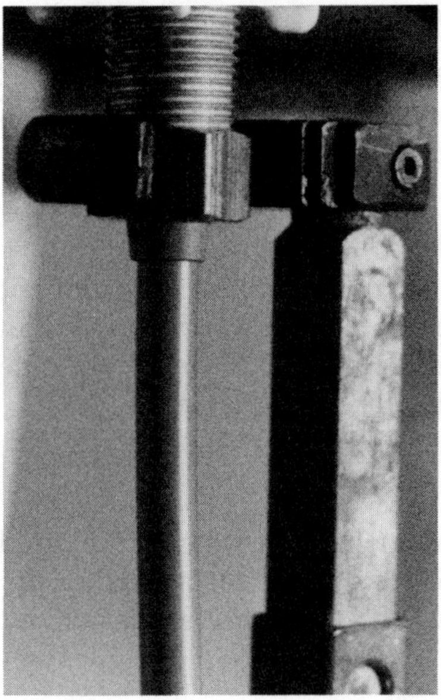

Figure 26.5 Basin wrench in use.

Troubleshooting Bathtub Drains

Troubleshooting bathtub drains is a lot like checking out lavatory drains. If you think of the tub waste and overflow as a lavatory pop-up, most of the techniques are the same.

When a tub won't drain properly, check for clogs in the tub drain. Some types of tub wastes are famous for making horrible hair balls. If the tub waste is not holding hair, consider partial drain blockages and blocked vents.

Bathtubs that won't drain at all generally require drain cleaning. However, before you go snaking the drain, check the tub waste for massive hair build-up.

If the tub does not hold water, look to the tub waste. If water is seeping past the drain plug, either adjust the tub waste or replace the drain plug.

General Principles

The general principles already discussed can be applied to various types of sinks, showers, and assorted plumbing fixtures. Remember to always use a process of elimination to find the problem you are required to fix.

Troubleshooting Electric Water Heaters

What should you look for when troubleshooting electric water heaters? Well, that depends on the nature of the complaint. Let's look at some of the most common problems and how to find them.

The water is too hot

There may be times when a customer will complain because the hot water is too hot. Correcting this problem should require no more than adjusting the water heater's thermostat (Fig. 26.6). However, in some cases the problem could be defective controls on the fixture.

Not enough hot water

A more common complaint is that of not enough hot water. This could mean that the thermostat needs to be adjusted, but it is more likely to indicate that the thermostat is defective or an element is burned out (Figs. 26.7 and 26.8).

No hot water

When no hot water is available, what could the problem be? There may be an electrical problem that is causing the water heater to malfunc-

Figure 26.6 Electric water heater thermostat.

Figure 26.7 Screw-in heating element for electric water heater.

Figure 26.8 Bolt-in heating element for electric water heater.

tion. The electrical power could have been turned off inadvertently. There is a chance that both elements have gone bad, or maybe the thermostat is shot. If no water—hot or cold—comes from the hot water side of faucets, check the valve on the inlet of the water heater to make sure the valve is open.

Water is collecting around the heater

What should you look for when water is collecting around the heater? The water could be coming from a hole in the water heater or from the drain. It is possible that a nipple or fitting has rusted threads that are allowing the water to leak. The water could even be coming from the relief valve.

What makes a relief valve discharge water?

Excessive temperature or pressure can cause a relief valve to blow off. If this happens, replace the relief valve. Sometimes an old relief valve just develops drips; this problem also calls for replacement of the valve.

Troubleshooting Gas Water Heaters

Troubleshooting gas water heaters is a little different than when working with electric water heaters. The potential problems with the relief valve are the same, but not all problems have the same cause with the two types of water heaters. There are no electrical connections or electrical heating elements in gas water heaters. Let's see how troubleshooting a gas water heater differs from checking out an electric water heater.

If a gas water heater is not producing hot water, check to see that the pilot light is lit. If necessary, put the controls through a manual test to see that the burner lights. If you can't get the burner to light, check to see if the gas valve is open. Should you be working with bottled gas, check to make sure there is gas available to the appliance.

The rest of your troubleshooting will be about the same as with electric water heaters.

Troubleshooting Submersible Water Pumps

Troubleshooting submersible water pumps can get complicated. You will need to know how to use an electrical meter to do this job.

What would cause a pump not to start?

The most common problems that prevent a pump from starting are no electrical power, incorrect voltage, bad electrical connections, and bad pressure switches.

What's wrong when a pump starts but cuts right back off?

If a pump starts but cuts right back off, there are several factors that could contribute to this problem. Electrical problems could include an improper fuse or circuit breaker, incorrect voltage, a bad control box, bad electrical connections, or a bad pressure switch. Other causes might be a pipe blockage, a seized pump, or a control box that is too hot.

What if the pump runs too often?

If the pump is cutting on too often, it could be the result of any of the following: a check valve that is stuck in an open position, a waterlogged pressure tank, a leak in the piping, a pressure tank that is too small, a hole in the pressure tank, or a pressure switch that needs to be adjusted.

Why would water pressure be low?

Low water pressure could result from any of the following: a pressure switch that is out of adjustment, a leak in the piping, incorrect voltage, or a bad pump.

The pump runs but doesn't produce much water

When the pump runs but doesn't produce much, if any, water, you should look for the following types of problems: a clogged strainer, a jammed impeller, a broken pump shaft, a check valve that is stuck in a closed position, a check valve that was installed backward, bad electrical wiring, a low water level in the well, a leak in the piping, or a bad pump.

Troubleshooting Jet Pumps

Troubleshooting jet pumps is a little easier since they are not hung in the well. Let's examine what you should look for in an ailing jet pump.

Low water pressure

If you've got low water pressure, there could be several causes for the problem. The strainer on the foot valve could be clogged. Sometimes just shaking the well pipe will clear the foot valve and correct the problem.

You might want to measure the lift demand on the pump. If the pump is forced to pull water too high, it can lose some of it's pressure. Other possibilities are a leak in the piping, a bad air charger, or a worn impeller hub.

Pump cycles on and off too often

If a jet pump cycles on and off too often, look for any of the following: a leak in the piping, a bad pressure switch, a bad air-volume control, a waterlogged or leaking pressure tank, or a pressure tank that is too small.

Pump does not cut off

If the pump does not cut off when its working pressure is reached, the problem is probably with the pressure switch. It may need to be adjusted or replaced.

Pump won't start

When the pump won't start, check the electrical power, connections, and voltage. A bad pressure switch could be the problem, or the motor might be bad. It is also possible that the motor contacts are stuck open or that the motor shaft is seized.

The pump runs but gives little water

When the pump runs but gives little water, you could have a pump that needs to be primed. If the foot valve is hanging above the water level of the well, you won't get any water, and a clogged strainer or a suction leak could cause this type of problem.

Well, we have covered the majority of the types of service calls you might get as a working plumber. We have also come to the end of the line. You have finished the book and should be ready to start your career in plumbing. I wish you the best and thank you for allowing me to contribute to your education.

Glossary

Accessible As it relates to plumbing, refers to a means of access. For example, a tub waste is considered accessible when there is an access panel that can be opened or removed to gain access to the tub waste.

Air-break In the drainage system, refers to an indirect waste procedure. The indirect waste enters a receptor through open air.

Air-gap (drainage) Refers to the vertical distance that waste travels through open air, between the waste pipe and the indirect-waste receptor.

Air-gap (potable water) The vertical distance water travels through open air, between the water source and the flood-level rim of its receptor or fixture.

Air gap (the device) A device used to connect the drainage of a dishwasher with the sanitary drainage system.

Antisiphon A device cannot be made to form a siphonic action.

Area drain A drain used to receive surface water, from grounds or parking areas, for example.

Aspirator A device used to create a vacuum, such as in a suction system for medical offices.

Backflow The backward flow of water or other liquids in the drainage or water system.

Backflow preventer A device used to prevent backflow.

Back-siphonage Essentially the same as backflow; it is the reverse flow of water or liquids in a pipe, caused by siphonic action.

Back-water valve A device installed on drainage systems to prevent a backflow from the main sewer into the building where the back-water valve is installed.

Ballcock An automatic fill device, most commonly found in toilet tanks. They supply a regulated amount of water, on demand, and then cut off when the water level reaches a desired height.

Branch A part of the plumbing system that is not a riser, main, or stack.

Branch interval A means of measurement for vertical waste or soil stacks. A branch interval is equal to each floor level or story in a building, but it is always at least 8 feet in height.

Branch vent A vent that connects individual vents with a vent stack or stack vent and can be used to connect multiple vents to a vent stack or stack vent.

Building drain The primary drainage pipe inside a building.

Building sewer The pipe extending from the building drain to the main sewer. Building sewers usually begin between 2 and 5 feet outside of a building's foundation.

Building trap A trap installed on the building drain to prevent air from circulating between the building drain and sewer.

Cistern A covered container used to store water, normally nonpotable water.

Code A set of regulations that govern the installation of plumbing.

Code officer An individual responsible for enforcing the code.

Combination waste and vent system A plumbing system in which few vertical vents are used. In these systems, the drainage pipes are often oversized to allow air circulation in the system.

Critical level The point where a vacuum breaker may be submerged, before backflow can occur.

Cross-connection A connection or situation that may allow the contents of separate pipes to comingle.

Developed length A method of measurement based on the total liner footage of all pipe and fittings.

Drain A pipe that conveys waste water or water-borne wastes.

Drainage system All plumbing that carries sewage, rain water, and other liquid wastes to a disposal site. A drainage system does not include public sewers or sewage treatment and disposal sites.

Existing work Work that was installed prior to the adoption of current code requirements.

Fixture supply The water supply between a fixture and a water distribution pipe.

Fixture unit A unit of measure assigned to fixtures for both drainage and water; used in pipe sizing.

Flood-level rim The point of a fixture where its contents will spill over the rim.

Grade The downward fall of a pipe.

Groundwork Plumbing installed below a finished grade or floor.

Hot water Water with a temperature of 110 degrees F or more.

House trap *See* Building trap.

Interceptor A device used to separate and retain substances not wanted in the sanitary drainage system.

Leader An exterior pipe used to carry storm water from a roof or gutter.

Local vent A vertical vent used with clinical sinks to transport vapors and odors to the outside air.

Main Any primary pipe for water service, distribution, or drainage.

Main sewer The public sewer.

Main vent The primary vent for a plumbing system.

Nonpotable water Water not safe for human consumption.

Offset A change in direction.

Open air The air outside of a building.

Pitch *See* Grade.

Plumbing inspector *See* Code officer.

Potable water Water safe for human consumption in drinking, cooking, and domestic uses.

Private sewage disposal system A sewage disposal system serving a private party. A septic system is an example of a private sewage disposal system.

Private water supply A water supply serving a private party, such as a well.

Readily accessible There is direct and immediate access to an object. If an access panel must be removed before the object can be accessed, the object is not readily accessible.

Rim The open edge of a fixture.

Riser A vertical pipe in the water distribution system that runs vertically for at least one story.

Rough-in The installation of plumbing in areas that will be concealed once the building is completed.

Sewage Liquid waste that contains animal or vegetable matter. The matter may be contained in solution or in suspension and may contain chemicals in solution.

Slope *See* Grade.

Stack A vertical pipe in the drainage system; may be a vent, soil pipe, or waste pipe.

Stack vent A vertical pipe that extends above the highest drainage point to vent the drainage system.

Storm water Rain water.

Sump vent A vertical vent that rises to vent a sump, such as in a sewer ejector sump.

Tempered water Water tempered to maintain a temperature between 85 and 110 degrees F.

Underground plumbing *See* Groundwork.

Vacuum A pressure that is less than the pressure produced by the atmosphere.

Vacuum breaker A device used to prevent a vacuum.

Vent stack A vertical pipe in the drainage system that acts only as a vent and does not receive the discharge of plumbing fixtures.

Waste A discharge from fixtures and equipment that does not contain fecal matter.

Water distribution system The collection of water pipes within a building that deliver water to fixtures and equipment.

Water hammer arrestor A device that defeats water hammer by absorbing pressure surges.

Water main The public water service.

Water service The pipe delivering water from a water source to the water distribution system of a building.

Well A water source from a hole in the ground.

Wet vent A pipe that receives the drainage from plumbing and does double duty by venting part of the plumbing system.

Yoke vent An upward connection from a soil or waste stack to a vent stack. Yoke vents are normally made with wye fittings to prevent pressure changes in the stack.

Index